Patrick Moore's Practical Astronomy Series

Other Titles in this Series

(Continued after Index)

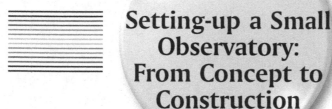

Setting-up a Small Observatory: From Concept to Construction

David Arditti

With a Foreword by

Patrick Moore

Springer

David Arditti BSc, MSc, PhD.
British Astronomical Association
d@davidarditti.co.uk

Library of Congress Control Number: 2007931612

Patrick Moore's Practical Astronomy Series ISSN 1617-7185

ISBN: 978-0-387-34521-5   e-ISBN: 978-0-387-68621-9

Printed on acid-free paper

9 8 7 6 5 4 3 2 1

springer.com

Contents

Foreword

Two books in this series, *Small Astronomical Observatories*, have been widely read and used. This new book by David Arditti gives the full story – how to choose a telescope, how to construct an observatory for it, and how to make the best of it when completed.

Astronomy is the best of all hobbies; it can take up as much time as you like – or as little. In any case, you will make many friends, and give yourself endless enjoyment. To be a proper "astronomer" you need an observatory, and this book tells you how to set about making one. Follow David Arditti's advice, and you will not regret it.

Here's to clear skies!

Patrick Moore

Author's Preface

Books on amateur observatories are quite rare, and most, if not all of them, in the past, have taken the form of collections of articles on particular observatories authored by their builders. The two books on observatories in this series already published, *Small Astronomical Observatories*, and *More Small Astronomical Observatories* are of this type. While useful, a danger of this approach is that it leaves gaps, and can be inconsistent in style and coverage. Discussing it with John Watson, the UK astronomy editor for Springer, we considered that it might be time for a more systematic and logical approach to be taken to the subject. Discussing it with other practical astronomers, there was also a feeling that the coverage of the subject in print had not kept pace with the technical developments that have transformed amateur astronomy, at least for some, in the last couple of decades, and that the examples in print now seemed rather old-fashioned.

This book is the result of these considerations. It became apparent to me, in writing it, that the real-life examples *are* very important and informative, and I have made these a major part of the book, while attempting to cover most of the possible approaches to the amateur observatory, from the simplest, cheapest, most portable and low-tech, to the most sophisticated and up-to-date, and covering both the DIY angle and the commercial. It is not possible to be completely comprehensive within a reasonable amount of space, and time to research, so I have also done my best to point readers to further sources of information on the subject, which will now come predominantly from the web. I have tried to write the book which would have been most helpful to *me*, when I first set about building an observatory. If it helps anybody else to achieve their goal of a practical and successful small observatory, I would be delighted.

Edgware, 2007

Acknowledgements

My thanks go to the other astronomers who have allowed me to feature their observatories in this book: to Bob Garner, Martin Mobberley, Olly Penrice, Dave Tyler, Norm Lewis, Es Reid, Richard Miles and Mike Morrison-Smith, for all their co-operation and help.

Thanks also to John Watson for his help and encouragement, to Mike Hardiman and Martin Andrews for advice in specific areas, and to Helen Vecht for reading the draft.

Errors and omissions are fully my responsibility.

David Arditti

Acknowledgments

Photo Credits

Figures: 1.3 Pop Outside.com, 1.4 top right, 9.43–9.50, 10.9 and 10.10 Norm Lewis, 1.6 Martin Lewis, 3.2 Dave Grennan, 3.3 Chris Go, 3.6 Alexanders Observatories, 3.9 and 3.10 SkyShed, 3.11 European Southern Observatory, 5.4 and 9.12–9.19 Olly Penrice, 9.4 Bob Garner, 9.8–9.11 Martin Mobberley, 9.40 Dave Tyler, 9.56 Es Reid, 9.57–9.59 Richard Miles, 9.60 Mike Morrison-Smith, 10.11 Ian Pass, 10.15 RobinCasady.com

CHAPTER ONE

Telescopes and Techniques

What is an Observatory?

When people think of an astronomical observatory, if they think of it at all, they usually think of a large domed building containing a telescope. And this is one type of observatory. But an astronomical observatory (as opposed to a bird observatory, or a meteorological observatory, which are not the topic of this book) is actually any site from which observations of celestial phenomena are made. It could be an open field, a tent, a garden, or any kind of building pressed into service for the purpose. Even important astronomical observatories have not been of the familiar pattern for very long. The observatories of the ancients tended to be towers, rooftops or high, open platforms, but there are few sources to tell us exactly what these looked like, or how they were used.

Figure 1.1 shows the greatest observatory of the pre-telescopic age, that of Tycho Brahe. This was a palace that he built for himself on the island of Hven (or Ven) in the Baltic, calling it Uraniborg, or "castle of the heavens". Observation seems to have been done both from the open balconies and the covered lookouts. By and large, in the pre-telescopic period, instruments were not protected from the weather – though Brahe built his second observatory below ground level to protect from the Baltic wind. Observatory buildings remained a rarity: Galileo made his great telescopic discoveries from a garden, and from windows in his house.

Even well into the telescopic age, specifically-designed observatory buildings were uncommon, and the domed observatory was unknown. Wren's design

Figure 1.1. Tycho Brahe's lavish observatory Uraniborg, in a contemporary etching.

for the Royal Observatory at Greenwich was one of the first attempts to solve a priori the architectural problems of accommodating telescopes in a building that protected them, but also allowed extensive access to the sky. It remains an elegant solution, with its tall windows and high spaces, well-suited to the long-focus refractors and transit instruments of the day. Later great observatories of the early telescopic age were often not enclosed at all, such as that of the Earl of Rosse at Birr, Ireland, with its huge telescope, for 72 years the largest in the world, suspended between massive walls, or the site of the totally impractical Craig Telescope, at one time the largest refractor in the world, perilously slung from a tower built for the purpose on Wandsworth Common in London.[1]

Domed observatories came into use from the mid-nineteenth century onwards. The first in England seems to have been the dome of the Northumberland equatorial telescope of the University of Cambridge, designed by Airy in 1835, but I am not sure there were not earlier examples. The design of the rotating dome with an openable slit running from the base to beyond the apex is a kind

---

[1]The fascinating story of this little-known heroic failure in the history of British astronomy has been researched by Greg Smye-Rumsby, and is documented on his webpage http://homepage.ntlworld.com/greg.smyerumsby/craig/

of optimum design for the astronomical observatory, in that it provides the most complete protection possible to the equipment while allowing complete access to the sky, and this is the reason why it has generally been the favoured solution of the professional astronomer, for whom budgetary constraints, and difficulty of engineering, are likely to be less of a consideration than they are for the average amateur. The classical dome shape seems to mirror the shape of the apparent celestial sphere itself; yet, the dome need not be of circular cross-section, and some major recent examples have not had the classical shape, but have had it simplified it to an easier-to-engineer cylindrical, or even cuboidal shape, as in the observatory of the highest-resolution telescope in the world, the Large Binocular Telescope (Fig. 1.2). Amateurs have been doing this kind of lateral thinking in their observatory-designing for a long time now.

The amateur of today who is considering some sort of permanent or temporary installation to make his observing more convenient or productive should bear in mind the wide variety of precedents and, after an examination of what his interests really are, what he really needs, and what would best suit his budget or his skills, come to some conclusion as to what equipment and structure would be most appropriate, if any at all. This book aims to help in making this choice.

Figure 1.2. The cuboidal "dome" of the large binocular telescope in Arizona.

One should not be afraid to think unconventionally. Is an observatory really needed at all, or could a small adaptation to the dwelling-house provide the answer? One of the greatest English amateurs of the 20th century, George Alcock, famous for his discovery of five comets and four novae, had a large plate-glass window put into his house, through which he trained his binoculars, to examine, with scrupulous care, for any changes, the star patterns he had so comprehensively memorised. On the other hand, knowledge of what you are doing is vital. Such a solution would be hopeless for telescopic viewing of the moon or planets at high magnifications. The window glass would not have the requisite optical flatness, and the temperature discrepancy on both sides of it would cause convecting currents of air that would ruin the "seeing", as astronomers call the state of steadiness or motion of the air in the light path of a telescope. An astronomer performing high-resolution work will have to tolerate the cold – unless, that is, he can devise some wired or wireless setup that allows him to observe indoors on a computer screen via some sort of electronic camera attached to a telescope outside. This type of arrangement will be discussed further later.

An observatory is essentially a practical item, though of course there is always room for aesthetic considerations. Many astronomers, or their family members, will not wish for a structure that is too visually intrusive or conspicuous, and there are related issues of security. Lateral thinking can lead to mistakes, as was mentioned in an earlier book in this series,[2] when one amateur conceived, with the object of concealing his installation as much as possible, the idea of a sunken observatory – and found that it immediately flooded because it was below the water table. (In fact, an observatory could be built below the water table, but it would have to be properly engineered. It would have to have a sealed concrete enclosure and below-floor pumping equipment.)

The best type of observatory for an observer of those objects best seen with the naked eye, or imaged with a fish-eye lens, that is, meteors, aurorae, noctilucent clouds, etc., is just a deckchair in a spot affording a good view of the sky, away from intrusive lights. Some sort of wind-break structure might be helpful for comfort, as it might be for using a portable telescope in the open. Such an observatory is complemented by a folding table, for star-charts and the like, and somewhere to rest binoculars or a torch.

It is possible to buy a moveable observatory, a tent, known as the Pop Outside Observatory (Fig. 1.3). This can be carried as a compact package, when folded, as a sort of back-pack. When unfolded, it has a 2 m (6 ft) square base, and viewing apertures can be opened in the sides. It is claimed it will accommodate up to a 20 cm (8 in.) telescope (though that would depend on focal length), and it can also be used to sleep two people. This could be a very attractive option for those making journeys to remote sites, or taking part in the "sky-camps" that have now become so popular on both sides of the Atlantic, with the object of obtaining less light-polluted views than can be had at home. Larger observing tents are also made by Kendrick Astro Instruments. I have seen reports that these perform poorly in windy situations. There is also a portable 3 m (10 ft) diameter

---

[2]More Small Astronomical Observatories, Ed. Patrick Moore, Springer, 2002.

Figure 1.3. The pop outside observatory.

domed observatory, constructed as an aluminium-framed PVC tent, called the AstroGazer, made by AstroGizmos (USA). As I am allergic to the discomforts involved in camping, I have not tested any of these products.

Of Telescopes and Allied Equipment

Before proceeding to the subject of the amateur observatory itself, I propose to use some space to discuss the issue of telescopes for various purposes. It is quite likely that the reader will already have his chosen instrument and be happy with it, in which case, he can move on to the "meat" of the book. There will be others, however, who are thinking of upgrading their equipment, and possibly of getting a larger or more sophisticated telescope in conjunction with setting up their own small observatory, and they may be interested in this overview.

Choosing a telescope can be a fraught subject. If one is buying, one is spending a substantial amount of money, and if making, one is expending a large amount of time and trouble (and probably quite a bit of money on tools and materials anyway – but the draw in this case is to end up with *exactly* what one wants, which is rarely or never achieved through purchase of a finished product). There was a time, and it was not that many years ago, when the hobby of amateur astronomy was almost synonymous with that of telescope-making, few being able to afford the price of a professionally-made instrument. But the cost of telescopes has fallen enormously, compared to average incomes, since then, and now the ATMer, the

Amateur Telescope Maker, is a rare creature. Whether the quality has gone up or down is a moot point. Clearly, much is available, in technological terms, that was not, say, 25 years ago, and such things as computer-controlled mountings, which were unknown then, have become almost standard. On the other hand, craftsmanship, build-quality, and basic fitness-for-purpose often seems to have taken a back-seat to cost-reduction.

This is most apparent in the mechanical build of commercial telescopes available today, and their mountings. Telescope showrooms, or adverts, are, unfortunately, full of instruments on mountings that are far too small and weak for them, and will certainly not allow satisfactory viewing at high powers, or with a slight breeze blowing on the tube. Accurate focussing, which depends on the telescope not moving or vibrating when the focuser is adjusted, will also be impossible. The stability of a mounting is of paramount importance – it is impossible to over-emphasise this fact, and, if it is purchased separately from the optics, which is usually the only way of getting a satisfactory result, expenditure on the mounting should be at least 50% of the total. Refinements such as GOTO and even basic electric drives should be considered secondary to stability.

The correct size of mounting for a telescope is often underestimated. This is mainly an issue with German equatorial mountings or GEMs (the type of mounting with a counterweight in a T-shaped arrangement balancing the weight of the telescope). Other types of mountings do not have so much potential for being undersized; fork mountings and altazimuth yokes have to fit round the tube, after all. Some experience of looking at systems and using them is required to become a good judge. To give an idea, some examples of telescopes on adequate GEMs, and some on inadequate ones, are compared in Fig. 1.4. If the mounting is over-specified for the telescope, so much the better – the option then exists of upgrading the OTA (Optical Tube Assembly) later if desired.

Twenty-five years ago, telescope-making was essentially a "cottage industry", localised, small-volume, craftsman-led, and close to the final consumer. Today, telescopes are mass-produced international brands, dominated by a very small number of companies, based mostly in the USA, Japan, Russia and China. These brands wield considerable commercial power, and it is debatable how much the amateur has benefited from this concentration of ownership. On the one hand, the technical sophistication of the products, and some aspects of their average quality, far exceeds what used to be available. On the other, it can be hard to get impartial advice, and there is limited choice. Further, an unfortunate fact needs to be pointed out to those seeking information on telescopes: reviews in the astronomical press are not necessarily objective. Magazines depend on the advertisements placed by the major manufacturers for too large a part of their revenue to be able to give fully objective assessments of products. The good points of commercial telescopes are talked-up, whereas glaring inadequacies in their design and manufacture go unmentioned, to be discovered later in the field by the hapless purchaser.

If the effect of what I have written here is to encourage the creation of more "ATMers", that would be no bad thing. There is a great deal of help, advice and support available through books, magazines, societies, and predominantly now, the internet, for anyone considering going down this route. With patience they will almost certainly get a better telescope than can be bought. It is not necessary

Figure 1.4. Observatory telescopes on German equatorial mounts. Top left: large enough, but poor design; top right: large enough, good design; lower left: marginally adequate, only for low-power work; lower right: too small.

to make every part of the telescope. The optics may, and in most cases will, be made by a professional maker. The tube-assembly can be made and fitted to a commercial mounting, or the mounting can be made as well. I will not say any more about this subject, but refer the reader to some of the excellent websites devoted to "ATMing" , listed in Appendix 2.

Telescope-making, however, is not for everyone. In terms of commercially-available products, the main lines currently available to the amateur are the following:

1. Newtonian reflectors of low focal ratio on simple altazimuth mountings – "Dobsonian" telescopes, as popularised by John Dobson of California in the 1980s. These give a maximum of aperture for a minimum of cash, but are not easily driven or guided, and so are unsuitable for high-magnification work

or for most types of imaging. They can, however, be very portable, being easily disassembled, to be put into a car and taken to a dark-sky site, and are appealing for those looking for simplicity of use and lack of technical complication.

2. Newtonian reflectors on German-pattern equatorial mountings. Once the mainstay of amateur astronomy, this design that has become far less popular in recent years. The telescope can be of short, medium or long focal ratio. The design is suitable for general purposes, for low and high magnification work, depending on focal ratio, and for driven and guided imaging and photography. It gives a versatile telescope with good value for aperture size, but limited portability in sizes above 150 mm (6 in.) aperture. Why has it fallen out of favour? It is hard to say.

   Newtonians are a certain amount of "trouble". They need care in collimation (getting and keeping the optics correctly aligned), and the mirrors need re-aluminising every 3–5 years, typically, depending on the environment they are kept in. But, also, Newtonians have not been surrounded by the commercially-generated advertising hype of other, more expensive, designs in recent times, such as short-focus refractors and Schmidt-Cassegrains. They have come, to some, to seem old-fashioned, cheapskate, or basic. And in the past, and even today, many of them were quite shoddily made. But all this is wrong. The Newtonian reflector is, in several meaningful optical respects, the most perfect telescope it is possible to construct. Newtonians require some understanding of the factors that affect their performance, such as thermal stability, in order to exploit to the full their potential, but they remain the mainstay of a great many of the more "serious" amateur observers – those who require maximum performance for their money, do not require a highly portable telescope, and are not too concerned with the vagaries of telescope fashion. Large Newtonians can also be mounted on massive fork-type equatorial mountings for greater stability, though this may mean counterweighting them at the mirror end. They can also used with the rarer types of custom mountings, such as the English equatorial or yoke.

3. Longer-focus refractors (above f8), usually on German equatorial mountings. These are usually only available in small apertures – up to 125 mm (5 in.) or thereabouts. They are portable, and very suitable for visual observation of the moon and planets, and for imaging. Their lightweight construction, combined with ungainly length in the larger sizes, may prove a problem for mounting and making stable ancillary equipment. One cannot not get far in visual observation of faint deep-sky objects with such telescopes, particularly in light-polluted environments, as the light-grasp is too small. They tend to be reliable and nearly maintenance-free.

4. Shorter-focus refractors – often described as "apochromatic", referring to the superior colour-correction (for the focal ratio) derived from using a triplet objective. They can be on German or altazimuth mounts, often with computer-control. They are either a means of spending a lot of money on limited light-gathering power and resolution, or of spending a truly enormous amount of money on only moderate light-gathering power and resolution. They are highly portable, flexible, and usable, but it is important to note the

actual image quality will never be quite as good as that of an equal aperture long-focus refractor of top quality, however expensive the short instrument. One pays for compactness in terms of compromise on the basic principles of refractive optics, that is the "long and the short" of it, but a point that the manufacturers usually will not stress.

In the smaller sizes they are currently very popular with deep-sky imagers, who use them to obtain wider-field images than can be obtained with any other type, apart from specialised wide-field cameras. In this application, they bridge the gap between telephoto lenses and other telescopes. Many imagers have one as an ancillary telescope mounted on a larger instrument, doing double-duty as a guide-scope and wide-field imaging scope.

5. Schmidt-Cassegrain telescopes. The Schmidt-Cassegrain design, a hybrid of the Schmidt camera and the classical Cassegrain telescope, using a fast (low focal ratio) spherical primary, a spherical secondary, and a tube-closing corrector plate, giving an overall focal ratio of usually 10 in a compact tube, was first marketed in the 1970s by the Celestron company of the USA, and subsequently imitated by others, notably Meade in their LX200 series telescopes (leading to lengthy patent disputes). It is not an exaggeration to say that, coming from nowhere, the design had come to dominate the world of amateur astronomical equipment worldwide by the 1990s, and it continues to do so. These telescopes are the most widely-available of all designs, popular in sizes from 100 mm (4 in.) to 350 mm (14 in.), and many amateurs will conclude now that they have almost no choice but to "join the club" and get one.

However, detailed consideration is advisable before going down this route. Firstly, SCTs are very expensive, for their aperture, compared with Newtonians – two to three times the price. They will not perform as well even if perfectly made (which they may well not be) because of fundamental limitations of the design. If portability is a major consideration, there is a strong argument for them, and even in the world of permanently-sited telescopes, there are good reasons for minimising tube length. The set-up will be easier to handle and the telescope will be lighter and easier to drive. The observatory, if there is one, can be smaller and cheaper and easier to use, and the telescope will be less prone to wind-vibration.

SCTs suffer from the problem of dewing of the corrector plate, particularly in damp maritime climates such as that of the UK, but there are remedies for this. These telescopes are also very collimation-sensitive, and do not keep their collimation well (due to the way the focusing system works, by moving the primary mirror). They therefore require skill in, and understanding of, collimation in order to get best results. They are flexible instruments, usable for wide-field imaging and viewing with the addition of a focal reducing lens, which decreases the effective focal ratio (EFL), or for fine planetary work with the addition of a Barlow lens or Powermate, both types of negative lens which increase the EFL.

SCTs are capable of giving superb results in the imaging sphere, as images from leading amateurs using them, such a Damian Peach and David Tyler, testify. Much of their advantage comes from the closed tube, which greatly reduces the tube currents that plague Newtonians and Cassegrains of other types, and effectively preserves the coating on the mirrors. They tend to be supplied on altazimuth fork mountings which may be converted to equatorials with the addition of a "wedge", adjustable to the latitude angle. The leading observers tend, however, to substitute high-quality German equatorials from other manufacturers. Celestron's latest lines have acknowledged this, with a shift to selling SCTs with computerised German mountings.

6. Maksutov-Cassegrain telescopes. Another hybrid design using a corrector plate closing off the front of the tube, "Maks" are similarly compact, and if anything, slightly more expensive for the aperture than SCTs. They are better-corrected for aberrations than SCTs, but tend to be only available in smaller sizes (e.g. Meade's ETX series). In the larger sizes they tend to be one-off builds costing a great deal. If the builder is well-known to, and trusted by the buyer, all is likely to be well, but to have one made to order by anyone else can lead to trouble. They are unlikely to be a serious consideration for the amateur observatory. The Maksutov-Cassegrains manufactured in England by Orion Optics (their OMC line) are open-tube instruments with a small correcting element inside the secondary, similar to the next type, but made in larger sizes.

7. Kletzov-Cassegrain telescopes. Made by the Russian firm Tal, in sizes up to 200 mm (8 in.), these are open-tube, compact Cassegrain-configuration telescopes which correct for aberrations using a small corrector lens in front of the secondary. They are unlikely to be used in an observatory.

8. Schmidt-Newtonian telescopes. Made mostly by Meade, these are fairly "fast" (low f ratio) telescopes. They appeal to those who like the sideways-on observing position of the Newtonian, but who want a compact tube free from air currents, and protected mirrors. They have smaller secondary obstructions than SCTs and therefore theoretically better performance on low-contrast planetary detail, though how much this counts in practice, particularly with modern imaging techniques, is open to doubt. They are more suited to imaging large deep-sky objects than SCTs, and have quite a flat field.

9. Maksutov-Newtonian telescopes. Similar to Schmidt-Newtonians, but normally of longer focal length, these are made by Intes Micro in Russia and other manufacturers. Again they are sold on their small secondary obstructions and reduced maintenance compared to standard Newtonians. Only available normally in smaller sizes, they are usually used with GEMs.

10. Dall-Kirkham Cassegrain telescopes. Only manufactured currently by Takahashi in Japan (their Mewlon series), the D-K design pre-dated the SCT, being invented by Horace Dall in England, between the wars, though not attracting much attention at the time. It is a variant on the "classical" Cassegrain design, substituting a spherical secondary for the hyperbolic figure of the classical Cassegrain design, and compensating with an under-corrected parabolic primary, which works very well for long EFLs. D-K Cassegrains give extremely good results in visual studies and imaging work on planets, and are fairly compact. They are unsuited to widefield work, as the field-edge suffers from significant coma (the comet-like spread of light from stars towards the field-edge, from which telescopes with fast primary mirrors in general suffer). They are available in sizes up to 300 mm (12 in.) on high-quality

Takahashi GEMs, but Celestron manufactures a 500 mm (20 in.) "modified D-K" telescope as well.

11. Ritchey-Chrétien Cassegrain telescopes. Made by a few specialised manufacturers, these telescopes are almost always made to order, and may be considered the "top of the range" for medium-to-wide field imaging and photography, having very flat (i.e. coming to focus all in one plane), coma-free fields. They are less-well suited to visual observing. Most large professional telescopes are built on the Ritchey-Chrétien pattern. They are extremely expensive, and mountings are bought separately, to taste. The "advanced Ritchey-Chrétien" telescopes now marketed by Meade are not true Ritchey-Chrétien telescopes, which have purely reflective optics, but a new hybrid, using a Schmidt-Cassegrain-like front corrector plate to compensate for a spherical primary. This is a good example of a manufacturer distorting or confusing the field of telescope nomenclature. Whether these telescopes have any real advantages over SCTs remains to be seen.

12. Cassegrain-Newtonian telescopes, in which a focuser is provided both at the side of the tube and the bottom, with a switchable Cassegrain secondary or Newtonian flat assembly used with a spider. This has the advantage that the telescope can be converted between a short-focus Newtonian instrument and a long-focus Cassegrain one, but the disadvantage that re-collimation is probably required each time the change is made, so it is not, in practice, likely to be done on a frequent basis. Some models are made by Takahashi.

13. The Cape Newise reflector, manufactured in England, is a new design, a variant on the Newtonian reflector, with a closed tube (though the closure is a flat optical window, not a corrector plate as in the Schmidt and Maksutov designs – and, incidentally, any other open-tube Newtonian or Cassegrain can be provided with an optical window rather than a spider, in principal, though for this to be done is rare). The Cape Newise uses, in addition, a lens that allows the primary focus to be further inside the tube than normal with a Newtonian, thus reducing the secondary obstruction, and a second lens that corrects for aberrations. It is another "catadioptric" design, like the Schmidt and Maksutov variants, using a fixed combination of reflective and refractive elements in its design. (The general idea of the Cape Newise has been around for a while, though not mass-manufactured, and a Cassegrain version of the same idea was made for a few iterations in England in the 1970s by the firm Astronomical Equipment. One of these "Dall-Kirkham-Dall Cassegrain" telescopes forms the main instrument in the author's observatory.) The Cape Newise is claimed to be a versatile telescope, good for high and low EFLs, and is supplied as an OTA only, the purchaser adding a mounting.

14. Solar telescopes. These are small refractors permanently fitted with a very narrow-band filter (technically known as an etalon) which transmits a specific wavelength of the solar radiation (usually hydrogen alpha) to reveal certain features not otherwise visible (such as prominences). Many amateur observatory-builders will want to include one of these in their setup as an ancillary instrument, possibly on the same mounting as the main telescope (Fig. 1.5). Larger refractors can be adapted for solar use by the addition of an etalon and an additional energy-rejecting front element, as can SCTs.

Figure 1.5. Small solar telescope mounted with a much larger telescope. It is attached with just a camera tripod screw.

This concludes our survey of telescopes currently on the mass-market. It is in no way exhaustive of the types of telescope that could be produced by the ATMer or commissioned from a custom maker. Important categories not represented are:

A) The classical Cassegrain telescope, still a good, versatile choice and possible to home-make, though much more difficult than a Newtonian.

B) The whole category of off-axis unobstructed reflector designs. (A very small unobstructed reflector is manufactured by US firm Orion, but generally they are ATM projects.) This is a complicated subject about which whole books have been written, and there is not space to go into it any further here. Such an excursion is for the experienced ATMer only, in search of something a little different. The claim is that eliminating the shadow of the secondary on the primary creates a level of image quality not available in conventional reflectors, but this is somewhat dubious. The disadvantages of the aberrations inherent in an off-axis approach, plus the large light losses incurred by the large number of reflections involved, probably create more deleterious effects than those eliminated by eliminating the tube obstruction. There are, however, a small number of serious enthusiasts who would certainly disagree with these statements, and their enthusiasm is not to be decried.

A decision about which type of telescope to go for will depend on the interests of the observer, as well as how permanent they want the set-up to be, how much they want to spend, and how much trouble they want to go to. As mentioned already, there is a whole category of observations that is best done with the naked-eye, such as those of aurorae, noctilucent clouds, and meteors. There is another category best performed with binoculars, which includes tracking artificial satellites, looking at open clusters and the Milky Way (and the Magellanic Clouds in the southern sky, if available from the location), and studying the brighter comets and variable stars and searching for novae. Some very serious binoculars are available, and the author has indeed heard of binoculars being made by ATMers of up to 300 mm (12 in.) aperture! (Such a contrivance would certainly be considered more a double-telescope than part of the conventional binocular category.)

Specialist suppliers of astronomical binoculars are listed in Appendix 2. The heavier binoculars, typically above 50 mm (2 in.) aperture, do require supporting on tripod-mountings for fatigue-free use. Simple adapters are available for fixing many binoculars to photographic tripods with standard camera-mounting threads, but such arrangements are not very satisfactory, because the binoculars are not mounted in the properly-balanced way that any telescope needs to be for smooth and controlled sweeping across the sky. The "parallelogram style" binocular mounts do not solve this problem either, relying on friction in the altitude hinge to keep them up, though they do move the observer clear of the tripod, which otherwise gets rather in the way in binocular astronomy. The best solution, however, is probably to have the binoculars looking down, rather than up, at a tiltable mirror, which is adjusted to view objects at different altitudes from a comfortable seated observing position. For even greater ease of use, the apparatus can be mounted on a turntable. Such systems are manufactured by Astro-Engineering in the UK, among others.

The subject of mountings becomes critical for telescopes of all sizes. Many people's early experiences of the hobby of astronomy have been marred by reasonable, small telescopes on foolish, badly designed, unstable mountings. The worst offenders are usually the tiny, undriven German equatorial mountings, supplied with very small refractors and Newtonians. For these telescopes, which must be used at fairly low powers, a simple, competent altazimuth mount would usually be better for the same money. The German equatorial mount (GEM) needs a high quality of engineering to work well; such is rarely found on budget, beginners' instruments.

The bigger the telescope, obviously, the larger and stronger the mounting needs to be, and in fact, since the weight of the telescope is the main factor, and this will increase roughly in proportion to the cube of its aperture, larger telescopes need very much stronger mountings than smaller ones. Unfortunately, these facts do not seem to be all that obvious to manufacturers, who want, typically, to produce a wide range of telescope sizes, but a very limited selection of mountings, for ease of production. This results in a tendency to market ranges of telescopes on a particular mounting, where not all the combinations are satisfactory. A good (bad) example of this is Celestron's "Advanced range" of SCTs. One mounting is marketed with four possible optical tube assemblies (OTAs), at 5, 8, 9.25 and 11 in. aperture (the last telescope is shown in Fig. 1.4 lower right). The mount is, if anything, unnecessarily large for the 5 in. It is about right for the 8 in., but the 9.25 in. is probably too heavy for it, and for the 11 in., it is quite inappropriate.

As a telescope increases in size, the dimensions of all parts of its structure need to increase proportionately, if parts of it are not to become too weak and flimsy. Once again, manufacturers, as a rule, seem not to understand this simple fact, which causes so many unnecessary problems to telescope owners. For example, one major manufacturer uses the same thickness of aluminium for all its telescope tubes, and for all its tube mounting rings. These tubes and rings are quite adequate for the 15 cm (6 in.) reflectors, but the increased mass of the optics and their mountings, along with the increased dimensions, means that the 25 and 30 cm (10 and 12 in.) telescopes in the range are wobbly, flimsy beasts, only useable at low powers, completely negating the benefits of the high-quality optics.

The search for lightness for transportation purposes, for those urban observers who wish to have a large telescope that they can take in a car to a dark-sky location, is no doubt a partial motivation for some of these problems, along with simple penny-pinching. Old telescopes tended to be much heavier for their size, which was in some ways good, and in others, bad.

One of the purposes of this book is to encourage observers to observe from their own backyards and gardens, wherever they live, if this is at all possible, rather than seeking "greener pastures" elsewhere. If we are to address and campaign on issues of light pollution, it ill-behoves amateur astronomers to be unnecessarily creating extra light pollution, noise pollution, fumes and general wasted energy transporting themselves and their equipment to remote sites, at least on a frequent basis. The restrictions imposed on observing by light-pollution are too easy to exaggerate. Much excellent observational work can be, and is, done by dedicated individuals working from inner-urban backyards and gardens, in small pools of darkness that either existed before, or that they have themselves created by good planning. Or they may have no real darkness, but confine themselves to work which can be done from a light-polluted location: observing and imaging bright objects, such as the Sun, Moon and planets, observing binary and variable stars, or carrying out filtered imaging of deep-sky objects. Probably the leading British visual variable star observer, Gary Poyner, observes from the city of Birmingham. There are some very successful deep-sky imagers, such as Tommi Worton and Bob Garner, who work in extremely light-polluted sites. Such city observers benefit from the time saved in moving, disassembling and assembling equipment, and travelling, so they actually use their time observing. An observatory, of course, optimises this time.

To return to the subject of mountings: some telescopes are heavy for their aperture, and some are light, and this will be a factor in determining how big a mounting is needed. The largest contributor to weight in reflectors and catadioptrics is normally the primary mirror. In sizes above 20 cm (8 in.), this weight becomes very significant, increasing at least with the square of the aperture. Such telescopes traditionally have mirrors 2.3–3.5 cm (1–1.4 in.) thick, which gives a large mass of glass. There has been a trend towards thinner mirrors recently (and lighter componentry generally), and some mirrors are made a conical shape to save weight. The support system in these cases has to be carefully designed to minimise mirror flexure. If a telescope is particularly long for its aperture, it will also need a stronger mounting, as it will be prone to vibration from wind and other sources. This applies to long-focus Newtonian reflectors, and to traditional large refractors (greater than 10 cm (4 in.) aperture).

A weak point in many telescope systems is the junction between the optical tube and the mounting. This fulcrum must not flex when the telescope is pushed, or there will be backlash, rendering it difficult to centre an object manually. Lack of rigidity at this point is a major contributor to problems of wind-vibration, and also shaky focusing – when, as soon as the observer touches the focusing adjustment, it becomes impossible to see the object to be focused, owing to the vibration imparted to the instrument.

Traditionally, with solid-wall telescopes (as opposed to skeleton or tubeless reflectors), the attachment to the mounting was by means of a saddle-plate and tube rings. With Newtonian reflectors, loosening the tube rings, or cradles, allowed the tube to be rotated, to obtain a convenient eyepiece position with equatorial mountings, and also allowed movement of the tube through the rings for longitudinal balancing. Tube rings have become less popular with manufacturers recently, with the shift to SCTs and other compact designs. These designs suffer from a problem consequent on their short tube lengths, that the range for longitudinal movement within tube rings is restricted, and a balance point may not be reached by this means. The usual solution is a type of dovetail-shaped metal bar attached to the tube, fitting into a recess on the top of the mounting, which can be slid up and down to balance the telescope (as in the bottom right of Fig. 1.4). Such arrangements, however, are vibration-prone, and inferior to the tube-ring system. Further, the adaptation of many commercially-available mountings to the dovetail system means that even telescopes with tube rings cannot be fitted to them optimally, as there is no strong, wide-based saddle-plate to transfer the load. These issues become less common at the upper end of the commercial mounting price range.

Mountings themselves constitute a large subject. In terms of mass-market products, the choice for observatory-sized telescopes is between simple, undriven, altazimuth (Dobsonian) mountings; computer-driven altazimuth fork mountings, which attempt to compensate simultaneously for change in altitude and azimuth; such mountings placed on an "equatorial wedge", which has the effect of converting them into equatorial, driven fork mountings; and German equatorial mountings (GEMs). For simple, un-technical, casual observing, and even for some less-casual observing modes, the minimalist approach of the Dobsonian has much to recommend it. There is almost nothing that can go wrong, which is refreshing in this age of technological complexity. The telescope also can fairly easily be disassembled, and transferred to another site. However, high-power observation, of planets, the Moon, close double-stars, globular clusters and planetary nebulae, while quite possible, will be a chore because of the constant pushing in two directions. And most observers wish to try imaging at some stage, for which such a telescope will be mostly unsuitable (though webcam imaging of the bright planets is possible with a Dobsonian telescope and a certain amount of skill, and excellent results have been so obtained (Fig. 1.6)).

To acquire a fork-mounted telescope, such as one of the Meade or Celestron SCTs, with the possibility of converting it to equatorial configuration later for imaging purposes, would seem to be a reasonable option, but, unfortunately, the forks, designed primarily for supporting the telescope in the upright altazimuth configuration, tend not to be really strong enough to give a good equatorial platform when combined with these manufacturers' standard wedges, which are, in themselves, not very adequate. The result can be a telescope that tracks the stars well enough, but vibrates like a tuning-fork. This may be good enough for some low-resolution types of imaging, and it will probably be possible to use the

Figure 1.6. An image of Jupiter taken by Martin Lewis using an undriven 22 cm (8.75 in.) home-built Dobsonian telescope and a Philips ToUcam webcam.

telescope as a driven platform for an ordinary camera, but it is not the system most accomplished astro-imagers use.

It is also possible to take images using the fork in the altaz. configuration, relying on the computer-control to do the tracking. This is a system that is normal in professional observatories. The drawbacks are: (i) that the computer control has to be very good, which it tends not to be on the amateur systems, and (ii), there is another phenomenon, that of field-rotation, which intrudes when one tries to track an object in the sky with an altaz. telescope. The telescope remains in the same orientation with respect to the horizon, but the celestial object does not, and gradually rotates during a long exposure. This is not an issue for short exposures of bright objects, but becomes one for nebulae and galaxies. It is possible to use software to de-rotate the image; nevertheless, most successful amateur imaging is done with purpose-built driven equatorial mountings, and the only ones available on the mass-market are GEMs.

With German mountings, the main choice now faced is between traditional simple mountings driven in RA, and maybe in dec. as well, and computer-controlled GOTO mountings. The difference is entirely one of price, as the computer-control is added on to basically pre-designed mountings by manufacturers, and can often be retro-fitted to a simple mount. But the observer needs to ask himself first if he needs the GOTO facility at all. He might be able to afford a heavier and more suitable mount for his telescope by forgoing it. If he is interested in observing primarily the Sun, Moon and bright planets, there will be little need for it (though one use for GOTO can be for finding Mercury and Venus in the daytime, when they are high up).

The computer systems used for controlling these telescopes tend, at the lower end of the price bracket, to be surprisingly crude compared to the computers we use for other purposes. For example, they don't keep track of the time and date when they are switched off, so demanding a wearisome ritual of resetting these parameters for every observing session, and answering other "silly questions". (They do, however, remember fixed parameters like latitude and longitude of the observing site.) The more expensive models avoid these problems, being much more user-friendly, and much more suitable for nightly use in an observatory, rather than merely an occasional portable foray to a dark-sky site, which is the most for which the cheap GOTOs might be tolerable. Again, the fast-slew drives on the cheaper units are very noisy, and potentially disturbing to the neighbours in the dead of night. They are also not terribly reliable. A fault-prone drive system is a curse, if you have to send your whole telescope or mounting away (sometimes to another continent) to be repaired. This can be the case with the level of integration of some of these units. You have nothing in the interim, not even a manually-operable telescope, to use. And the failure rate of some of these units seems to be very high. Reliability seems to be enhanced by minimising the use of the fastest slew rates.

On the other hand, the better GOTO systems do work very well, and can be a boon for locating elusive objects from light-polluted sites (that is almost everywhere anyone lives, now). They save time, allowing more objects to be observed in a session, and are easier to use and potentially more accurate than old-fashioned setting-circles (which are, in themselves, not to be despised). They reduce the need to spend time learning the constellations and star-patterns in order to find objects in the sky. But of course some would say that is the essence, and the most enjoyable part, of amateur astronomy or stargazing – the chase being as interesting as the capture. It is certainly a feature that distinguishes amateur from professional astronomy, as traditionally practised. Few professional astronomers "know" the sky in the way that most amateurs do. However, GOTO is certainly here to stay, and many serious amateurs working on faint objects would not now do without it, for the increased productivity it affords; equally, others try it, and find there are no advantages in it for their type of astronomy.

Computerised mountings have been built by amateurs, though such a task is beyond most. However, the scope for home-built mountings of a more basic functionality is very large, and these can easily be made far superior to most commercial offerings in terms of stability. For very large amateur telescopes (reflectors in excess of 35 cm (14 in.) aperture) the prices of adequate commercially-made equatorial mountings become enormous, as very few are made; in this case, the self-build option, or a special commission from a custom maker, becomes more likely. There are at least six possible designs that can be explored (Fig. 1.7).

The German equatorial mount is so-named because the prototype was built by Joseph von Fraunhofer for the Dorpat refractor in Estonia in 1820. The large majority of "serious" amateur telescopes today are on German equatorial mountings. Adequate size and strength has already been stressed in connection with GEMs. This is particularly important because of the doubly-cantilevered nature of the GEM: the polar axis is hug over from the pier, and the dec. axis is hung over from that. Even some big and massive examples of the type can

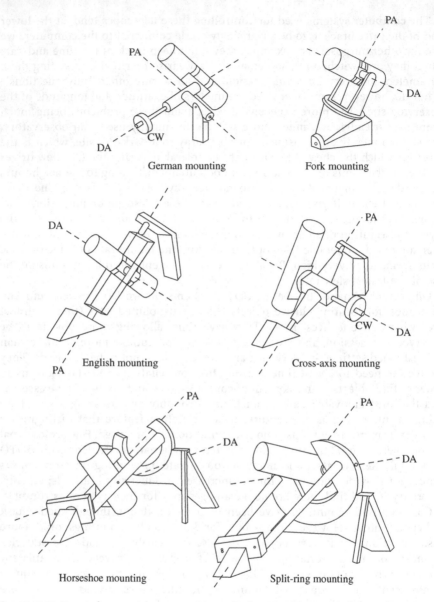

Figure 1.7. Types of equatorial mounting PA = polar (or RA axis), DA = declination axis, CW = counterweight.

have surprisingly poor performance if not well-designed. The first mounting in Fig. 1.4 is a case in point. It has too much mechanical leverage operating on the long shafts, without adequately-large thrust-absorbing surfaces. It is thus too resonant. The best GEM designs have RA and dec. assemblies which taper away from wide thrust surfaces; the RA section should be larger than the dec. section, and the whole should be compact, as in the second example of Fig. 1.4, not widely cantilevered, as in the first example. Adequately rigid connection between telescope and mount is a particular issue with GEMs. Another common weak-point in GEMs lies at the junction between the base and the polar axis assembly, the hinge at which adjustment for latitude is made. There is a lot of moment acting on this junction, which causes vibration in mounts too weak for their load. The two upper examples in Fig. 1.4 have substantial, well-designed mount bases.

The main problem with the GEM operationally is the requirement for the mount to be "normalised", soon after an object has been tracked across the meridian. This means reversing the positions of the telescope and counterweight, so the telescope can continue to rise above the level of the polar axis, rather than falling and soon colliding with the pier. Exactly how much travel is available with the telescope below the level of the counterweight depends on the detailed design. Increasing this range makes for more convenience in tracking objects, but makes it more difficult to maintain rigidity, as the telescope's centre of mass has to be displaced further from the polar axis.

A considerable advantage of the fork mounting, and the reason why many observers prefer it over the GEM, is the avoidance of the disruption caused by normalising. The fork design is good for low f-ratio reflectors and catadioptrics, but has to be strongly constructed, even more so than a GEM of similar size, because the telescope's centre of mass is well away from the RA bearings. The length of the fork arms, or tines, has to be sufficient for the telescope and all equipment at the eye-end, with Cassegrain configuration telescopes, to swing between them. If the construction is too light, it can result in a resonance exactly as occurs in a tuning fork. The fork mount can work very well if properly designed "from the ground up". But it may be noted that the result of placing a typical SCT altaz. fork on a wedge is a long way from the idealised, properly-supported fork design illustrated in Fig. 1.7.

The design of the English mounting, first devised by George Biddell Airy for the Duke of Northumberland's telescope at Cambridge, England, is intrinsically very stable, though it does need a lot of room, and demands a large and totally permanent observatory, because of the requirement for two support piers. In this design, the weight of the telescope is carried between the polar axis bearings, as well as between the dec. axis bearings as with the fork. The mounting is not cantilevered in either direction, but is totally and strongly supported (though the upper pier is usually slightly cantilevered to improve clearances). The celestial pole, and the area below it, is obscured, but this is not a serious disadvantage, as the area below is always presented above at other times, and the polar area, is, in any case, difficult to access with all equatorial designs (as is the zenith with altaz. designs).

The English mount is sometimes known as the "yoke" mounting, because the telescope is carried in a yoke, and yet other authorities describe it as a "polar frame". It can be constructed in many different ways, and can even be of reinforced wood (as was Airy's original design). The axes are reduced virtually to two pairs of bearings. Some versions raise the line of the dec. bearings slightly above the frame or yoke. The telescope can then look at the pole, but not below it, and counterweighting is then required on the other side of the yoke. This complication is probably not worth the gain. In its usual form, the English mounting is not counterweighted, and is not normalised.

Another variant is also sometimes known as the English mounting, but this is referred to here as the cross-axis type. The telescope in this case lies on a short counterweighted declination axis which allows it to see over the polar bearing to the pole. This type is more like a German mounting with the top of the polar axis supported as well as the bottom, while the English mounting is essentially an extended, supported fork. In these types, the upper and lower supports of the polar axis must be very solid, and can be of metal, wood or masonry, but they will be heavy and permanent. The cross-axis mounting shares with the German the inconvenience of having to be normalised near the meridian. A cross-axis telescope will generally collide with the ground if not normalised. English and cross-axis mountings are rare on amateur telescopes, but more common on large professional instruments.

A new type of equatorial mount was designed by American amateur astronomer H. Paige Bailey in the 1930s. This is the "horseshoe" mount. Russell W. Porter more or less copied this design for the 5 m (200 in.) Mount Palomar reflector (later named the Hale Telescope, for a long time after World War II, the largest telescope in the world), and most sources wrongly credit the invention to Porter. The horseshoe design combines the solidity of the English mount with the ability to see the whole sky. It does this by extending the top polar bearing of the English mounting into an interrupted circular track (the horseshoe), through which the telescope can look when pointed north.

Porter does seem to have been the originator of another variant on the same idea, known as the split-ring mounting. This moves the declination axis into line with the horseshoe or split-ring, for even greater rigidity. In this variant, the yoke of the English mounting has been designed out, along with its possible resonance, the horseshoe or split-ring becomes the upper part of a large conical bearing, and the telescope passes through the ring in crossing the celestial equator. The horseshoe and split-ring mountings (which are sometimes confused in the literature, because of their similarity) are mechanically the most perfect of the equatorial designs, as they combine the cantilever-free principle with a view of the whole sky. The engineering of these designs is challenging with amateur resources, and there are few amateur examples, though there are some. (Small versions of the split-ring have been made in plywood.)

Amateurs have recently tended to copy more the modern professional trend of fabricating large altazimuth mounts with computer-controlled dual-axis drives for large telescopes. Few of these, however, have been very successful, and few would want to invest so much effort in a telescope-making project. The method of providing a drive for any of the equatorial designs is a large subject which can only be touched on here. Usually it is by means of a worm gear and toothed wheel set converting the small torque (turning force), at high rate, provided by an electric motor, to the large torque, at a low rate, required by the telescope. The worm and wheel are very accurately made components, which only a very well-equipped and skilful amateur can fabricate, so these will probably be bought.

Though there are other driving methods which are sometimes used with large telescopes, worm and wheel arrangements are almost universal, and are the invariable rule on commercial mountings (though they are often hidden). They suffer from a fundamental problem, that of periodic error. Because the worm will not be precisely circular in cross-section, the rate at which its edge drives the wheel attached to the polar, or RA, axis will vary slightly in a regular pattern, repeated after one turn of the worm, its "period". This will manifest itself as a drift of the object being studied, eastwards and then westwards, in the field of view, or on the imaging chip, with that period. Periodic error is the bane of long-exposure imaging, and can never be eliminated completely, though there are various strategies for compensating, either manually or automatically.

The periodic error (PE) of a mount is one of its most important parameters, and should be investigated before purchase, particularly if acquiring a mounting with a view to undertaking imaging work. Normally the cheapest mounts have the largest PE. The larger the worm and wheel, the smaller the PE should be, though this does not always follow, as it also depends on the accuracy of manufacture. Some large old worm and wheel drives were very inaccurate. Damage and wear makes these things worse. It used often to be recommended that the wheel diameter should equal the telescope aperture. In few systems is this criterion met, and neither does there seem to be any particular logic behind it, as the main issue is the symmetry of the worm, which bears no relation to the size of the wheel, though a larger telescope will require a larger mount, and a larger mount should be larger in all its components. Modern manufacturing has probably rendered this criterion obsolete. PE during imaging can be compensated for by computer methods, but it is always best to keep it to a minimum mechanically.

The worm and wheel, or whatever form of gearing is employed, will be driven by an electric motor. This may have other gearing attached, but its accuracy will be much less critical because it is not the final stage of rate-reduction. Telescopes used to be driven by falling weights or clockwork drives, but there is no need to discuss these here. Virtually all driven telescopes working today have electric drive motors, of which there are four main types: mains-voltage AC synchronous motors, low voltage DC motors, low-voltage stepper motors, and low-voltage servos. Older telescopes tended to have synchronous motors, and quite a lot are still in use. They have the disadvantage of requiring high voltages for their operation, which is potentially dangerous in the damp atmosphere of an observatory, so good earthing is required and, ideally, isolation from the mains. They will only give a small range of drive-rates, which can be obtained by varying the frequency of the supply, which is done with a variable frequency oscillator, normally transforming up from a low-voltage DC source. Hence they cannot be used for fast slewing, and are generally incompatible with GOTO systems, and modern computerised drive-rate correction systems. They have an advantage of being free from the vibration which sometimes can be a problem with stepper motors, if their resonance happens to coincide with that of the mounting.

DC motors are rare, but found on some Meade telescopes. They are difficult to regulate to a precise and predictable rate, and are not recommended. Most modern telescope systems use microprocessor-controlled 12 V stepper or servo motors. With stepper motors, the microprocessor board feeds pulses to the motor windings, which causes the speed to vary between rapid slewing from one part of the sky to another and precise tracking rate. Counting of the pulses sent allows the circuitry to predict how for the telescope has moved from an initial reference point, which allows it to keep track of position, and makes possible GOTO. There need be no high voltage near the telescope, and the level of user-control is high, if all works correctly. The telescope does not need to be pushed at all – it can be completely controlled from a handset or external computers.

Stepper motors can induce vibration in the mounting because of the discrete steps they use in their operation, and increasingly they are being superseded in GOTO systems by servos. These have been used on high-end mounts for some time, and are now being found at the budget end of the market as well. Servo systems use a feedback loop for control, whereby a miniature opto-mechanical system measures the precise rate of the motor, and a microprocessor uses that information to regulate the motor and determine how far the telescope has moved across the sky. GOTO mount stepper or servo motors are generally locked on to the axes in tracking and slewing, so that the telescope cannot be moved by hand, except by unlocking them. The systems are tolerant of poor balancing of the telescope, though this should still be avoided, as it causes unnecessary wear. Old mounts driven by synchronous motors tended to have loose slip-clutch arrangements between the drives and the axes, so had to be rather precisely balanced to work. On the other hand, this meant it was impossible to damage the gears and motors by forcing the axes, or by running the telescope into part of the observatory, which is much more of a hazard with powerful modern drives.

A strength of the latest GOTO systems is their integration with available computer software. A computer can be connected so that it is synchronised with the mount, continually querying the mount position encoders to establish the pointing direction of the telescope, which can be displayed on-screen, either as co-ordinates, or as a target on a planetarium-type display. The software can allow the user to point and click on an object in the display, and instruct the telescope to automatically slew to that position in the sky.

The most basic telescope drives using stepper motors do not have GOTO operation, but it can often be added later as an extra. An alternative to GOTO is "push-to", that is, a system of rotary encoders fitted to the mounting shafts separate from the drives, feeding positional information to a computer or dedicated display (the latter often known as "digital setting-circles"). "Push-to" can work with planetarium programs to show the observer graphically where the telescope, manually slewed, is pointing at any moment, and can be integrated with synchronous, DC, or hand-driven telescopes, including Dobsonians and other altaz. instruments.

While a drive in RA is more or less essential for imaging, and greatly aids high-power visual work, a drive in declination is always an optional extra, and many observers get on quite happily making declination adjustments manually. Often the declination adjustment is by a hand-operated worm and wheel while the telescope is driven in RA. The hand-control, however, usually tends to be in the wrong place at any particular time, so this is not very effective. A dec. drive operated from the same handset as the RA drive is preferable, and it must have a reversible motor (synchronous motors are not normally reversible). The dec. gearwheel is often much smaller than the RA gear because less precision is needed in dec. adjustment. On an altaz mount, both gears would be the same size.

All in all, it must be said that to make a satisfactory equatorial mounting is more difficult than to make a satisfactory telescope, in that it needs more skill and more equipment. Many who make reflectors will fit them to a commercial mounting, and may indeed also buy the mirrors and some other components. There is a large and fruitful area between the complete "homebrew" solution, which few will be inclined to seek, and the purely commercial offering. This area is the realm of adaptation of bits of equipment from various sources into a personalised instrument, and many examples of this will be found in this book. Most amateurs, at some time or another, probably make their own adaptations and modifications to manufactured instruments in order to make them more effective for their purposes. For all the efforts of the manufacturers, it is not quite possible, and probably never will be, to buy complete satisfaction and success "off the peg" in amateur astronomy.

Visual, Film or Electronic Observing?

The greatest change that has come upon amateur astronomy in recent times has been the shift from visual observing and recording with pen, pencil, charcoal, etc., to the recording of observations electronically, for various purposes, including imaging, positional measurement (astrometry), and brightness measurement (photometry). The use of film photography is still common in some types of observing work, notably the recording of meteors, but it is undoubtedly "on the way out". The CCD (charge-coupled device) and CMOS (complementary metal oxide semiconductor) chips and their associated electronics and digital processing techniques can now do all that film and visual recording could do in the past, and much more. This revolution was affecting professional astronomy as early as the 1970s, and by the 1990s it had hit the amateur world with full force as well.

The planner of an observatory would do well to consider at an early stage what imaging methods and equipment he is likely to be using, as this will influence the design of the project, and interact with the choice of telescope-type and general areas of astronomy in which he is most interested. If he only anticipates ever being a "casual observer", then the observatory need only accommodate the telescope and the observer, along with a few shelves for eyepieces, and maybe a chair. The equipment associated with electronic imaging however, does expand the space that is required. Again, if the telescope, mount and observatory is to be made fully automatic, suitable for remote operation, then a very small observatory could be used, with the electronics transferred to another building, possibly the dwelling-house. But this is a specialised kind of project that few will attempt. Most people will want to observe from their observatory, sometimes visually, sometimes using cameras of various sorts, and sometimes using computers, and most will want sufficient space to accommodate guest observers from time to time.

The ubiquity of the laptop or desktop computer in amateur astronomy is now well-established. Even if not used for collecting observational data, it is likely to be used to display the sky on a planetarium program, to guide or aim the telescope, to keep an observing diary, or perform other functions. Most people will want to think about computer equipment in the planning stages of an observatory. A laptop can be used on a small shelf or table, which maybe can be folded away when the observatory is required to accommodate more people. A permanent desktop setup with a monitor will be cheaper, more expandable and probably more powerful flexible and ergonomic, but bulkier, and less easy to get out of the way. The advent of cheap LCD displays has alleviated the bulk issue with desktop computers, however. One can be hung on a wall and take up almost no observing space.

If using a desktop computer, there will need to be some means of taking the data away, probably to another computer indoors, for processing – in other words, either some kind of removable disk of sufficient capacity, or a wired or wireless local network. Few people will want to spend hours processing their observations in a cold observatory. The files generated by modern imaging methods are very large, often measured in gigabytes. A laptop gets around this, as it can just be taken inside with all the original data on its hard disk. Also, a laptop is likely to be electrically safer and more reliable in the damp conditions often encountered in an observatory.

One method of using a telescope that has become popular in recent years involves the use of a video camera to take and amplify the image, which is then fed to a TV, and possibly to a video or DVD recorder. Using this method, fainter objects and finer detail can potentially be seen than with the eye at the eyepiece. This is a boon particularly to those contending with light-polluted skies. The way is also open for the observer to operate from a heated room instead of having to sit with the telescope at night air temperature. The cameras are sometimes standard video and "night vision" surveillance cameras, slightly modified for the purpose, or they may be purpose-made, such as the Astrovid and Watec cameras. Standard camcorders have also been used to good effect, particularly on planets. All these cameras can be made to work best if the lens on the front can be removed to allow direct coupling to the telescope. There is a wealth of useful information on the internet (see Appendix 2 for links) placed by amateurs who have developed these methods, and indeed web-based groups, such as the famous QCUIAG (QuickCam and Unconventional Imaging Astronomy Group), have been the principal means of communication between amateurs involved in developing these techniques, linking those who have an interest in writing software, those who understand the hardware, and those who principally observe and record.

The video camera methods are particularly suited to recording precisely timed phenomena such as occultations of stars by minor planets, and a significant area of amateur-professional collaborative research at present is the study of the shapes and sizes of these objects by this method, which yields far higher accuracy than traditional methods using the eye and a stopwatch.

Another, related area of technical development on the amateur scene in recent years has been the use of webcams to record detail on the bright solar system objects: the Sun, Moon, Venus, Mars, Jupiter and Saturn. These tiny, mass-produced cameras were designed for purposes such as video-conferencing over computer links, and when the first amateurs started trying to record astronomical images with them, most were highly sceptical that anything would be achieved. However, the technique has been developed spectacularly, achieving resolution undreamed of from amateur equipment only 10 years ago, even surpassing the best results from professional telescopes of that era, and approaching the resolution achieved by some space-probes. The names of the leading practitioners of this art and science have become well-known to readers of the journals and magazines, through their spectacular published images: Damian Peach and Dave Tyler in the UK, Don Parker in the USA, Christopher Go and Isao Miyazaki in the Far East, to name but a few.

A webcam is just a camera containing a small CCD chip and associated electronics, which interfaces directly with a computer, usually through USB 1, USB 2, or FireWire (IEE 1394) connections, and takes a large number of frames in quick succession. Webcams are not all that light-sensitive, hence their use mostly on bright objects, but their strength lies in their high rate of data capture. Used on telescopes with very large effective focal lengths (thus projecting the image of a tiny planet onto the detector chip at a large scale), they take images at a rate of maybe 5–40 frames per second, which are saved to the computer's hard disk in real time. Because the resolution of moderate-size telescopes (above 10 cm or 4 in. aperture) is largely determined by the constantly fluctuating state of the atmosphere, known as the astronomical seeing, webcams will normally capture a mix of good, sharp images, and poor, blurred ones, when run over a few seconds or minutes. Computing techniques are then used to filter out the poor images, and construct a resultant "stacked" image (in other words, an average image) from the good ones.

The more frames that are stacked, in general, the better the results, under even mediocre seeing. The signal-to-noise ratio increases proportionately to the square root of the number of images averaged. The best-known piece of software for performing this image stacking operation is a piece of freeware called *Registax*, developed by Cor Berrevoets, but there are others. Further image manipulation, in various programs, is then normally carried out in order to achieve the finished result.

Traditional photography never achieved good result on planets, because of the tendency for seeing fluctuations to blur out individual exposures, and in fact visual observation and drawing remained the best method of recording planetary detail as late as the mid 1990s (short of sending a probe up there). The trained human eye and brain is remarkably good at detecting and fixing detail in a badly fluctuating image – performing, in fact, a similar computational averaging procedure to that performed by software such as *Registax*. However, the webcam-computer combination can now obtain better results under any conditions than the eye and hand, and these results are more objective and reliable – though they are not automatic, and as much skill and dedication as ever is still required on the part of the observer, to exploit conditions and opportunities to the maximum, and extract the most real data out of the signal recorded. The skill involved in observing is never reduced by technology, but it does change its form.

An advantage of webcams is that they are quite cheap, and little extra equipment other than a telescope and computer is really required. Many people buy the computer for other purposes, anyway. Webcams cannot, in their normal form, be used for recording faint objects, such as nebulae, clusters and galaxies, because they will not permit long exposures. However, ingenious amateurs have discovered that they can be modified so that they can expose for as long as desired, and can be used for recording deep-sky objects. (The limitation on the exposure length is the thermal noise that is intrinsic to the CCD detector, and eventually fogs out the image.) Software has been developed to control these modified webcams, of which the best-known item is *K3CCDTools*, by Peter Katreniak. The modified cameras can be bought from certain suppliers. They are not so good as true astronomical CCD cameras, but are much cheaper. The main difference between the latter and the former is that the true astronomical CCD cameras are cooled by some method to well below the ambient temperature, to greatly increase sensitivity by decreasing thermal noise. However, fair results may be obtained on the brighter and more compact deep-sky objects using air-cooled modified webcams, such as the Atik Instruments ATK 1HS camera, in colder locations – there are some advantages to living in a cold climate! (See Fig. 1.8).

The cooled CCD cameras are expensive – anticipate spending as much on one as on a high-quality telescope. They are capable of amazing results, as any trawl through an astronomy magazine, or the websites of imagers, will show. The major manufacturers are Starlight Xpress (UK) and SBIG (Santa Barbara Imaging Group) (USA). At the top-price end they merge into the equipment used by professional astronomers. For work such as surveys of distant galaxies for new supernovae explosions, or imaging faint comets, or showing the spectacular details of nebulae imaged in narrow optical wavebands using filters, they have no competition. The cooling allows for long exposures, of many minutes, but, of course, long exposures require very accurate tracking. There is no point embarking on long-exposure imaging if your mount cannot be made to track accurately, with either a very low periodic error (PE), or a mechanism that satisfactorily compensates for PE and other causes of image-drift.

Figure 1.8. Messier 1, the Crab Nebula, imaged by the author with an ATK 1HS modified webcam and a 20 cm (8 in.) driven (but unguided) SCT.

The two possible compensatory mechanisms are as PE training, by which the computer control is taught by the observer what the mechanical errors are over a period, and then automatically compensates for them, and auto-guiding, where a separate camera, or guide chip, detects drift due to PE and other causes and, operating through a computer and an interface to the mount, adjusts the drives in real time to compensate. Traditional manual guiding using a subsidiary telescope of long focal length (a guidescope) and crosswires is also possible. It should be noted that even given a mechanically perfect mount, with negligible PE (and some mounts do approach this), guiding is still necessary for very long exposures at typical telescopic focal lengths because of changing atmospheric refraction with changing altitude, telescope flexure, and because, in the case of comets and other solar system bodies, the target does not move at sidereal rate, it moves additionally with respect to the stars.

The requirements of PE correction and guiding are eased if image scale is reduced, as then the drift is slower and less noticeable in a given length of exposure. Additionally, for diffuse objects, such as nebulae and galaxies, the exposure required goes down as the light in the image becomes more concentrated, at smaller effective focal lengths. Hence the recent popularity of smaller, shorter focal length telescopes, particularly small apochromatic (triplet objective) refractors, on high-quality oversized computerised mounts, for use with CCD imagers. These produce spectacular results on large, diffuse nebulae, such as the North America Nebula, which are hardly obtainable any other way. Larger telescopes, unless they are specially constructed astrographs, devised specifically for widefield imaging, do not give a wide enough field to do such objects justice. A complicated imaging setup based on a small telescope probably requires an observatory almost as much as a simpler large telescope does, as the setup time can be impractical if all the equipment, cameras, computers, etc. have to be brought outside every time and connected up. The wonder of an observatory is that everything can be left connected, ready and waiting, in situ.

Image-scale is also reduced by using a larger CCD chip combined with high-sensitivity "binning mode", so that individual elements in the detector have their signals combined in groups, thus reducing resolution, but increasing signal. This is analogous to using a high ISO film in photography. CCD cameras with larger chips and binning capabilities become rapidly more expensive. The chips in "entry-level" CCD cameras are less an 8 mm (0.3 in.) across, and it is important to realise this when trying to predict what result will be achieved with such a camera, compared to that from a film camera using film 35 mm across. There is not space here to go in detail into the techniques of CCD imaging, but there are excellent published accounts.[3]

A cheaper option than the cooled CCD, particularly viable for wider-field work, remains the film SLR camera, connected to a wide-field telescope by T couplings, or used with a telephoto lens (effectively a small-aperture wide-field telescope), a standard or a wide-angle lens, with the camera mounted "piggyback" on the main telescope, using its drives and guiding. Fine shots of the Milky Way, the

---

[3]See, for example, *Digital Astrophotography: The State of the Art* by David Ratledge, in this series.

Magellanic Clouds, constellations and open clusters, such as the Pleiades, can be taken in this way, but a dark, unpolluted sky is really required. The orange glow of sodium lights, so ubiquitous in towns now, quickly fogs film, though it can be partially filtered out with light-pollution filters. Of course, exploitation of film photography to its potential requires the user to have their own darkroom and processing rig. Sensible results cannot be expected from commercial processing labs, whose staff will have no idea of what the astrophotographer is trying to show. Because of this, and the rapidly developing digital competition, film astrophotography is being practiced less and less.

The direct alternative to the traditional high-quality single lens reflex (SLR) film camera is the digital SLR (DSLR) camera. These cameras are based on CMOS detectors, which have higher thermal noise and hence lower sensitivity than CCD chips, but they can be manufactured in large sizes much more cheaply. The size of the detector chip in many fairly inexpensive digital cameras now approaches the old 35 mm film standard.

Most consumer-level digital cameras, though excellent in many other respects, have one major disadvantage for astronomy – the lens cannot be removed. It is possible to use these cameras with a telescope, utilising some method of attaching the camera so that it "looks into" an eyepiece, but such arrangements, even involving specially-manufactured accessories, tend to be slightly rickety, unstable and unsatisfactory. Fairly good images of lunar and solar features are the best that can be expected from this so-called "afocal" method of photography. A level of manual control on the camera is desirable, so the automatic settings can be overridden, and also a timed or cable-operated shutter release.

The DSLR camera is far more suitable than these consumer-level cameras for all astronomical purposes. The lens can be unscrewed, allowing direct coupling to the telescope using suitable adaptors (usually two adaptor rings, one having the universal T-thread, and the other adapting this to the particular make of camera.) The camera can be used at prime focus, in other words, using no optics other than the primary (and secondary in a reflector or catadioptric scope) for a low image scale, or it can be used with Barlow lenses or eyepieces in the light path (Barlow projection and eyepiece projection) using suitable accessories, increasing the effective focal length (EFL), or it can be combined with a focal reducer, a lens that is essentially the opposite of a Barlow, in that it decreases the EFL, to reduce image scale further beyond the prime focus scale. Focal reducers can be used with modified webcams and CCD cameras as well, but these generally will still give smaller fields than the DSLR due to the smaller detector size. Focal reducers are popular with SCTs, generally reducing them from f10 to f6.3 or f3.3, but can be used with other telescopes as well with suitable couplings.

DSLRs can be used with telescopes to record fine wider-field views of open clusters, extensive nebulae, star clouds, the Andromeda Galaxy, etc., but they are rapidly fogged-out by light pollution (and moonlight), limiting exposure in badly-polluted areas to only a minute or so. Various narrowband and broadband filters can mitigate the pollution problem, as they can with CCD cameras. Post-processing in an image manipulation program can also be used to reduce the effects. (Later in this book, I will discuss the observatory and methods of Bob Garner, who achieves amazing deep sky imaging results from one of the most unsuitable, light-polluted sites imaginable.) One strategy is to take several shorter exposures and stack them, using similar software to that used for processing webcam images. This reduces the build-up of background signal, and also alleviates the requirement for precise tracking. Again, this is a method that can also be used with modified webcams and CCD cameras, but not, so easily, with film.

The DSLRs are rapidly getting better and cheaper, with larger detectors and greater sensitivity. The most popular makes with astronomers are Canon and Nikon. The most popular models are undoubtedly the Canon EOS 300D and EOS 350D at present, but no doubt improved successors will more prevalent before this book has been long in print. DSLRs allow rapid downloading of the images to a computer for perfecting in a graphics program, such as *Photoshop*, *PHOTO-PAINT* or *Astroart*, or the images can be recorded direct to a computer. The exposures can also be controlled by computer with suitable connections and software.

DSLRs, like SLRs, can be used alternatively with standard, telephoto, or wide-angle lenses in telescope-piggyback mode to record wider views of the sky (as indeed can most CCD units). Taken together, the CCD and CMOS cameras have ushered in a new world of amateur astronomical imaging (arguably the word photography should no longer be used), with easy processing, free from the vagaries and mess of chemicals and darkrooms, and independent of commercial photo processors who have no idea of astronomical photography. Some astronomers will continue just to watch the sky and wish for no more, and some will continue to draw and paint, but most will wish to use the new technologies to record both things they can and cannot see. The exciting possibilities opened up by the new cameras will no doubt continue to advance into the foreseeable future.

A Summary of the Options

We will conclude this introductory chapter, before getting on to the business of planning the observatory itself, with a summary of some of the equipment and accommodation options available for different types of astronomy.

For some purposes, notably, wide-field photography or imaging and observations of astro-atmospheric phenomena such as meteors, aurorae and noctilucent clouds, and visual observation using binoculars, including such useful work as the visual monitoring of brighter variable stars, there is no real need for an observatory, as only lightweight or simple equipment is involved, though some measure of shelter from wind may be desirable for the observer and equipment.

Study of the bright objects in the solar system, either visually or electronically, requires a high resolution telescope, ideally above 15 cm (6 in.) aperture and at least f6, even better, f10 and above. Such a telescope and associated equipment is not going to be easily portable, and many observing opportunities will be lost if it is not permanently set up in an observatory, both through the actual time the setup takes, and the reluctance of any sensible person to spend half an hour setting-up only to find at the end of that time that the clouds have rolled over again. Telescopes used for lunar and planetary work will generally be long-focus Newtonians and the various forms of catadioptric scopes and Cassegrains. Refractors above 15 cm aperture are rare and very expensive.

Visual observation of the deep sky is easy and convenient with a short-focus Newtonian on a Dobsonian mount, and such an instrument need not be in an observatory, as it is fairly easy to disassemble and store, but it will be limited basically to visual observation at low powers. For the larger Dobsonian, many will still wish for the convenience of an observatory, with the option of occasionally dismantling the telescope to take on holiday to a darker sky, or to one of the "astro camps" which are now such popular gatherings. In an observatory, the Dobsonian may be more easily developed into a more sophisticated computerised observing platform, if desired.

Imaging of the deep sky can be carried out with a relatively small telescope and a good mounting, but the quantity of ancillary equipment, cabling and computers required will soon make temporary setting-up a chore, unless one is lucky enough to live in a rather predictable climate. Accurate polar alignment of an equatorial mounting will have to be performed each time, and it will never be as accurate as it would be in an observatory, in which it can be gradually perfected over time. In addition, other adjustments, such as dual-axis balancing, will have to be made each time, and moreover, reflectors and catadioptrics lose precise collimation when they are moved, and this has to be reset more frequently in a portable set-up than in an observatory (the small refractor scores here, as not generally needing collimation).

If imaging of the smaller deep-sky objects, such as planetary nebulae and globular clusters, is to be attempted, we are more in a high-resolution regime, as with planets, best results requiring large apertures and very heavy and stable telescopes and mountings, and again a permanent site becomes a necessity. A portable telescope, even on an excellent tripod, can never be as stable as one on a properly-built permanent support. It is not totally necessary to have an observatory structure around a permanent telescope site, and this possibility will be dealt with in Chapter 3. For our purposes, an observatory is any fixed observing site, whether or not walled or roofed-in any way.

The study of double stars, either casually, or for purposes of precise measurement, which has not been mentioned so far, also requires, optimally, a long focal length instrument, which can be of any type, so long as the effective focal ratio can be taken into the region of f20 and above with amplifying lenses (Barlows, Powermates or relay lenses). Traditional long-focus refractors, and the various types of Cassegrain, score well here, and again, a very solid, permanent base is almost essential.

The reader should now be in a position to decide whether an observatory is an appropriate solution for his or her needs. In most cases it will be. It is now necessary to consider exactly where it should go, and how to go about making it acceptable to everybody else.

CHAPTER TWO

The Observatory Site

Selecting the Best Site

The prospective observatory builder may have very little choice over his or her observatory site, or a lot, depending on how much space he has available. One question that often comes up is: can part of the dwelling house be used as, or extended into, an observatory? I would say this is possible, but not to be recommended. Observatories have been built on roofs of houses, or as extensions to them, but I would advise against it. The main problem this creates is thermal instability. A heated building will, inevitably, greatly interfere with the operation of a telescope, which needs to be at the temperature of the outside air to within a fraction of a degree Celsius in order to avoid currents of air inside and outside it destroying the resolution completely. An additional problem would be one of stability. It would be very hard to stabilise a large telescope sufficiently, so that there was no transmission of vibrations from the rest of the building, were it to be mounted high up in a house roof. Once again, it has been done, and great efforts have been made by such builders to provide a solid foundation high in the air, but I do not recommend it. It is bound to be a very expensive operation, which will certainly require the scrutiny of the planning authorities, and is unlikely to give satisfactory results except maybe for very casual and occasional viewing. And what is the point in going to a lot of trouble and expense for that?

A rooftop may seem to be an ideal location, with good, unobstructed views, but in fact, in most locations, this situation will also expose the observatory to the maximum of wind and the maximum of local artificial lighting. In general the region of the sky below 20° altitude is not much use for observing anything – it suffers too much from atmospheric absorption and refraction and turbulence. Obstruction of the 20° nearest the horizon is not much of a loss, and will be a good thing in many locations, where there are streetlights and house lights to contend with. True, interesting phenomena often do occur low down, but they tend to be the things at which there is little point aiming a large telescope. When the spectacular Comet McNaught appeared in the sky in northern temperate latitudes in early 2007, very low down in the west at sunset, I found I could not see it from my observatory, and observed it with binoculars, telephoto lenses and a small telescope from the upper floor of the house. Telescopic images of this object showed no more than images with good telephoto lenses. The same had occurred a few months earlier for a partial eclipse of the Moon, visible at moonrise in the east.

The place for an observatory is at ground level. But what about adapting an existing garage or shed, in its current location? This may be viable, if it is not too close to the house, or if it lies to the south of the house (in the northern hemisphere). Most observation is performed on objects close to the meridian, when they are at their highest, lying due south, or at any rate to the south-west or south-east. Mercury and Venus and many comets need to be observed when they are close to the western or eastern horizons, and other special events may also need to be observed in other parts of the sky, but basically, in the northern hemisphere, telescopes look either upwards, or south, and in the southern hemisphere, they look north. Therefore, if the site can be arranged with the nearest heated buildings in the opposite direction to this, that will be optimal. It is also desirable for the site to be as far from heated buildings as possible, within reason. Obviously, if one is dealing with a large tract of land, one will not want to have too long a walk to the observatory, but few people will be in this position. Most people will find the optimum location is more or less at the bottom of their back garden.

An adaptation of a garage will be examined in Chapter 9. There are certain disadvantages to this course. Most garages have foundations which are essentially a large, wide, but not very deep slab of concrete. Unless this is excavated and rebuilt for the telescope, which is difficult, this kind of a foundation tends to pick up surface-wave vibrations in the ground. This is a problem near main roads and railways. Elsewhere, it may not be a problem. Other, smaller, outbuildings might also be adapted, but in most cases, the adaptation will be so much work and expense that replacing them with an a priori designed observatory will make more sense. Again, their existing foundations are unlikely to be of the right type for a substantial telescope, and, with wooden sheds, their construction will probably be too light to allow them to work as an observatory without major reinforcement – in which case, why not start with a clean sheet and design a shelter with ideal dimensions?

I will assume for the rest of this discussion that we are selecting a site a priori, that has no existing constructions on it. It will be apparent from what has been said that (in the northern hemisphere) the most important horizon is the southern, and the least, the northern. The site thus should not be badly obstructed by trees or buildings to the south, though in urban areas it is likely that some obstruction will have to be tolerated. Obviously, one makes the best of what one has available. The location of local artificial light sources needs to be considered next, and whether, if they cannot be avoided, they can be screened in some way. This might be with some construction like a raised fence, or trellis, and by using plants creatively. It could be a wooden or hardboard screen, or it could be a long-term project to cultivate fast-growing plants, like conifers, to the correct height. A screen will tend to catch the wind, and might blow away – something natural is preferable. Coniferous plants are ideal. Other trees, of course, lose their leaves in the winter and so do not provide much of a screen in the main observing season.

Light pollution is clearly the most important problem facing the 21st century astronomer, and here it might be worthwhile saying something more about dealing with light pollution at source. You do not necessarily have to put up with the fact that your observing site is flooded with artificial light, and you do not necessarily have to move. Sources of artificial light can be categorised into private, i.e. un-curtained windows, security floodlights, etc., and public, i.e. streetlights and illumination of sports pitches and the like.

It may well be possible to persuade your neighbours to change their habits and even spend some money on making sure that their lighting arrangements have minimal impact on your observing. Invite them round and show them the night sky through a telescope. Most people are sympathetic to harmless, environmentally-friendly hobbies like astronomy. The most frequent problem is outdoor "security" lighting, and the enormously excessive 500 W "rotweiler" floodlights, either switched on continuously, or movement-sensitive, and continually triggered erroneously by animals, the astronomer, or other causes. To make things worse, these are often installed pointing at an angle too close to the horizontal, flooding adjacent land with light while not illuminating the owner's house or ground close to it, and providing, in fact, no real security at all, but most likely masking the presence of any intruder, and making their activities easier. In many cases a householder, when shown the problem, will agree either to modify the lamp, or even buy a more "astronomer-friendly" security light of much lower wattage, with better-directed beam. They will incidentally save on their electricity bill in this outcome.

If all else fails, the possibility exists of taking legal action to reduce domestic and industrial light nuisance, but this is not a course to be entered upon lightly, and negotiation is always preferable. A new provision was passed into English law under the Environment and Clean Neighbourhoods Act 2006, which allows an individual in England and Wales to complain to a local authority about intrusive light from premises, and gives the local authority powers to deal with it as a "statutory nuisance", within the same legal framework as is used to deal with noise and other environmental problems. However, this law has not yet, at the time of writing, been tested, and it appears that local authority officers, unsure as to its meaning and scope, and as to how bad the light intrusion has to be, are reluctant to use it. The possibility also exists in both English and US jurisdictions of taking out a private prosecutions on grounds of light nuisance (the term "light trespass" is often used in the United States), but these claims appear only to have succeeded in the past where the complaint has stressed the health effects of the intrusive lighting (lost sleep), rather than the effect on conditions for astronomy.

The new act in England and Wales does not apply to public lighting such as streetlights. Nevertheless, the situation with streetlights can sometimes be improved. A local authority may well agree to modernise one or more of the streetlamps adjacent to your property, replacing the luminaires (the light-emitting glass bowl part of the lamps) with less polluting "full cut-off" ones, which do not direct any illumination above the horizontal. This may or may not be helpful. Full cut-off luminaires make more difference to light pollution some way from the source (skyglow) than immediately adjacent (light intrusion). Alternatively, the local authority may agree to paint part of the existing luminaire black to reduce light intrusion, or to install some kind of screen.

Factories and businesses may agree to turn their lights off after a certain time, or reduce them (again saving themselves money in the process), if the point is made. The most difficult cases tend to be sports pitches, golf courses, and rail and other transport depots, which are often very inefficiently and intrusively illuminated, and are explicitly excluded from the 2006 English legislation.

In the United States, the situation is complicated, as there is no federal law on light pollution, but there is a certain amount of legislation at the state, county and municipal levels. New Mexico has the most progressive and comprehensive statute limiting light pollution, requiring most outdoor lighting to be shielded, and Arizona, Connecticut, Maine, Michigan and Texas also have some useful provisions in their state law. A great deal of further information and advice on this whole subject can be obtained in the UK from the Campaign for Dark Skies, and internationally, from the International Dark-Sky Association, which has much information of relevance to North America (see Appendix 2 for links).

An observatory site ideally should be level, though some slope can be accommodated. The best substances to look out over are grass, plants and foliage. Concrete, brick and stone, and bare earth, all tend to absorb sunlight in the day and re-emit it as infrared radiation during the night, causing local warming and air currents that damage seeing. Plants emit very little heat at night. The country-dweller may not be an ideal situation, seeing-wise, if his observatory is surrounded by ploughed fields. Again, keep in mind that the most important direction is the south (in the northern hemisphere).

A too-open site, even if there are no problems from artificial lighting, may suffer from strong winds. The prevailing wind direction should be established. Inland in the British Isles, it is normally from the west, but it might not be in your particular location. If the observatory is too exposed, firstly, it could be damaged by the wind, and, secondly, the telescope might suffer too much from wind vibration. Wind-breaks might be considered, and, as with the limiting of unwanted light, something natural is to be preferred: a hedge, or other plants which can be kept down to a reasonable height. Obviously, these will take time to grow, so some temporary solution may be appropriate in the meantime.

The site must be large enough, and here we had best pause to consider how large an observatory actually need be.

How Large Should It Be?

Apart from the design of the observatory, which we will discuss in the next chapter, this is the most fraught question for the observatory planner. There will be a natural desire to prevent costs running away, to avoid a too-heavy construction which will be difficult to use, to aim for a reasonably time-limited project, and to avoid creating a too-intrusive "blot on the landscape". We will deal with the question of "camouflage" in the next section. These factors will often conspire, with the first time observatory-planner, to lead him to conceive something that will prove far too small to be really usable. Remember, the observatory may be a once-in-a-lifetime enterprise, so it is worth doing it properly, and making it a sensible, usable size.

Commercially-made domes are available in sizes from about 2 m (6 ft) diameter upwards. They become far more expensive with increasing diameter, so most amateurs who go for this option tend to opt for the smallest size. This is a mistake. A circular observatory, for general purposes, needs to be *at the very least* 3.3 m (10 ft) in diameter, and 4 m (12 ft) is far preferable. I am neglecting here the possibility of a completely automated setup, which is never used for direct visual observations. In this case, a very small observatory can suffice. However, the proportion of amateurs who want this will be very small, and they will all be advanced practitioners.

Square or rectangular observatories are usually self-built or commissioned from a builder or carpenter, either general or specialised. An observatory with corners allows much more space than a circular one occupying the same plot. For this reason, a 2.4 m (8 ft) square observatory is just large enough for a medium-size telescope (which I define as less than 25 cm (10 in.) aperture in a short-tube instrument). Again, larger will prove much better in the long term.

Consider that in the centre of the observatory floor will be the telescope pier or stand, on which will rest the mounting (Fig. 2.1). Let us suppose the telescope to be 1.3 m (4 ft) long. It will be necessary for people to be able to get round this, but its length will not be distributed equally on both sides of the mounting. In a reflector, the mirror end will be very heavy compared to the open end, so the tube will be asymmetrical in the observatory. Note also that, with a German mounting, the whole tube is offset from the centre of the mount. You can make the measurements and calculations on your instrument for yourself, but, probably, you have to allow for an obstruction of the space by the telescope tube of about 1–1.3 m (3–4 ft) when it is pointing at low altitudes. You can see from the diagram that distance A is almost equal to the length of the tube.

This gives a minimum radius for the observatory of 1.2–1.5 m (4–5 ft) in order for the observer just to be able to squeeze round. But there will need to be furnishings in the observatory. There will need to be a chair, or stepladder, shelves, may be a cupboard, and almost certainly a desk. This adds at the very least 0.3 m (1 ft) onto this required radius. It is easy to see, if the logistics are thus simply considered, how the diameter of a circular observatory simply has to be between 3.2 and 3.8 m (10–12 ft) *at least*, depending on details of the telescope and usage. The commercial fibreglass and plastic domes of 6–8 ft inside diameter really are quite pointless, and must be discounted from consideration immediately. Going down the path of an undersized observatory will lead to long-term frustration and dissatisfaction. It is a long-term investment, so it is worth doing at a sensible scale.

As we have noted, a rectangular or square observatory is more space-efficient (and it could still have circular dome). Furnishings can be placed in the corners, so the basic minimum side length here would be 2.4–3 m (8–10 ft), depending

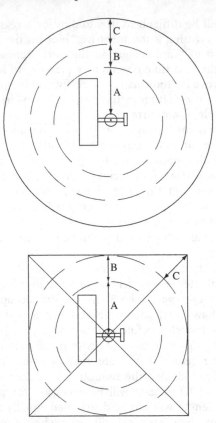

Figure 2.1. Clearances and space in a circular and a square observatory of similar "spaciousness": A is the clearance required by the telescope, B is the space needed by the observer to get round, C is the space required for furnishings and storage.

on telescope and usage. But what is reasonably comfortable for one observer can easily become overcrowded if a few people – family members, friends, or a gathering of local amateur astronomers – wish to experience the observatory. A square 8 ft observatory is overcrowded with three people in it, and, having built one, I have often wished, on these occasions, I had built it bigger. Think ahead – you will be glad you did.

Although these calculations and diagrams are based on a fairly long telescope, they are very conservative, and the calculations will not be found to be very different with a short-tube SCT or other compact telescope, because, in fact, these telescopes need a long dew-shield more or less permanently fitted to them for use in most climates. Their compactness is something of an illusion.

There is a noticeable tendency for American observatory constructors to "think bigger" than British, perhaps a consequence of the latter living on such a crowded island. British observatory-planners are often faced with the long, narrow gardens of old terraced housing, in which a decent-sized observatory will often occupy the whole width of the land. To persuade other family members to accept that percentage diminution of the available space may be difficult. However, the suggestions in the next section may help. Two common designs, to be described in Chapter 3, the run-off shed and the run-off roof observatory, effectively require twice as much space as the building itself, as the building, or its roof, move away from the telescope en bloc. However, this space is not completely lost to other usage. It can be used as lawn, deck, terrace, or planting space, as will be seen.

Finally, under the heading of space considerations, it must be borne in mind that most astronomers, sooner or later, wish for a larger, more powerful telescope. This may well require a stronger, more massive mounting. It will be as well to build in redundancy in all directions in the observatory, in terms of diameter, height, foundation depth, and mounting pier strength, if at all possible, within restrictions of finance and available land, so that, when that happy day comes, the larger telescope can be installed without a complete, costly and disruptive, rebuilding of the observatory.

Environmental Considerations

We will now consider in more detail the subject of how to make the observatory acceptable to others. Variables are the type, size and height of the observatory, its materials, colour, and general intrusiveness. It is always good policy to keep neighbours informed and consulted on what you are proposing, quite apart from following statutory planning procedures. Initially, of course, the issue will be gaining acceptance from other members of your household.

The most visually intrusive type of observatory is the white dome made of fibreglass or other material. There are good reasons for it being white: it will reflect most sunlight, preventing the interior from becoming too hot in the daytime, which will mean that the equipment inside does not takes too long to cool down at night. White domes are in some ways ideal, but a compromise on colour may be necessary to please neighbours and other family members. Green domes are quite popular, and a light shade of green should not absorb too much heat. Another approach is to make the dome reflectively metallic (probably aluminium sheet or aluminium paint), so that it reflects the sky colour. This should be quite good thermally, but experiences indicate that it is not so cool a finish as gloss white paint.

As the next chapter will show, there are plenty of observatory designs which look more like normal garden buildings, and blend into urban and suburban gardens better than domes. Wooden constructions are very effective from this point of view, particularly if they use a shiplap or weatherboarding method of construction, similar to most sheds and garden fencing.

Siting an observatory in the middle of a lawn may give it the best horizons, but it may not be what others particularly want to see. Consider putting it behind a partial screen of trees or bushes that will isolate it visually from the house, and also block stray light from house windows as seen from the observatory. A large garden, or a long narrow one, can be divided up and screened in such a way that the visual intrusiveness of an observatory can be kept to a minimum.

Figure 2.2 shows my observatory as an example, as viewed from my house. Because of the screen of trees and bushes that divides the garden crossways,

Figure 2.2. My observatory as seen from the house in summer.

Figure 2.3. The trellis separating my two telescope sheds, with tomatoes growing. Behind is the wooden deck on which the run-off shed (to left) moves. To the right is the run-off roof observatory, with its front roof rail at the top of the trellis.

the observatory buildings (there are three of them: a run-off roof observatory, a run-off shed, and a shed for remote telescope control) are almost completely invisible from the house in the summer. What can be seen is merely some boarding, which could be a fence or any normal garden structure. This is the creation of what garden designers call an "indefinite boundary". On one glance, the central shrubby bed appears to be the limit of the garden; then, on a closer look, the fruit tress and lawn behind this can be seen, and the wooden structures behind these appear to be the limit. In fact, they are not the limit either. A hedge 1.2 m (4 ft) behind them forms the actual boundary. The garden is 25 m (80 ft) long in total, but the idea could be scaled down for shorter gardens.

There is a trellis between the fronts of the two observatory buildings, allowing plants to grow in this space (Fig. 2.3). This is necessary because the space behind is used both for the movement of the roof of the run-off roof observatory, and for the movement of the walls of the run-off shed. This space has been made to look like a deliberate garden design feature, and the rails and platform necessary for the observatory operation look like part of the same design, being all wood. But in fact, none of this is particularly visible from the surrounding properties, because of the number of low trees and hedges around. The trees and hedges in the garden have been kept down to 4–6 m (13–20 ft), a level that supplies screening of surrounding lit windows, as seen from the observatory, and also a level of protection from the wind, but does not raise the horizon too far. Fruit trees should be kept to this level, or lower, in any case: they will produce more fruit at a level at which it can be conveniently harvested. The plan of the garden is shown in Fig. 2.4.

These observatory buildings have been kept particularly low, below 2.4 m (8 ft) above the lawn level. There are disadvantages as well as advantages to this. They cannot be stood up in by a normal-height person unless they are in operation. Domes score here, as, although they normally need a particularly low door, they can usually be stood up in at all times, which is handy for working on the equipment inside them on rainy days.

There will be an infinite number of ways of incorporating an observatory into a garden or backyard, and each case will be different, depending on the dimensions of the plot and of the observatory, and on what is already there. Hence there is probably no point in giving further examples, but to encourage each person to use their imagination.

Other garden buildings around an observatory, since they will not be heated, will probably not affect conditions much, so long as they do not obstruct the view. Even greenhouses cool down very quickly at night, and so are unlikely to be a problem. Brick and concrete buildings are the ones that cause most thermal problems. My telescope control shed is wooden, and can be heated, but it is placed to the north–west of the telescopes, so will not often be looked over by them.

Camouflaging an observatory not only helps placate neighbours and family members, it makes for security, as the structure is less likely to attract the attention of criminals. It will not be obvious that there is valuable equipment inside if the observatory looks like an (albeit slightly odd) garden shed.

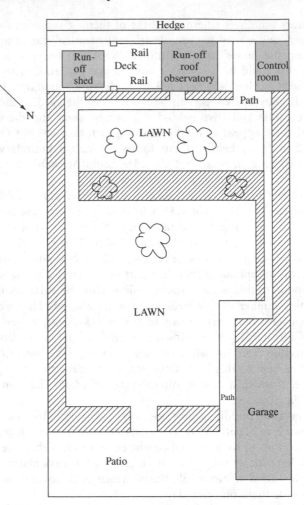

Figure 2.4. Plan of my garden and observatory (hatched areas are planted).

Statutory Planning Requirements and Building Codes

Statutory planning requirements and building regulations, or codes, differ between different jurisdictions. Something will be said here about the procedures in the UK and USA, for initial guidance only. It is essential you contact your local planning authority to find out if you will need permission for your observatory, what conditions you will have to meet, and how much you will have to pay for permission, and possible inspection by a planner.

In the UK, *planning permission* is required for any new building more than 3 m high, or 4 m high if it has a ridged roof. Observatories may or may not require planning permission, by this criterion, depending on design. Domes are covered by the 3 m limit, as they are not ridged. Whether the building is fixed to the ground or not is irrelevant to planning, contrary to statements in some astronomy books. If the observatory is below 3 m in height, it will still require permission if either: it results in more than half the land around the original house being covered by buildings, or, if it is closer to the highway than the original house (unless it is more than 20 m from the highway), or if it is less than 5 m from the house. A building less than 5 m from the house and of volume more than 10 m³ is always regarded as an extension to the house, and definitely needs permission. So an observatory close to a house is almost certainly covered. There are additional regulations in conservation areas and national parks in the UK.

An application for planning permission in the UK will be directed to the planning department of the local authority. A fee will be payable to submit an application, and a plan and a description of what is proposed will be required. Planning officials will check whether building regulations are being satisfied, and may come to inspect the work when it is complete, to check the structure has been built as agreed.

In the USA, there are two distinct bodies of regulations which need to be satisfied by new buildings: *zoning ordinances*, published by each city, and *building codes*, published on a state by state basis. The zoning ordinances establish the type of building use of an area, and the building codes establish the technical requirements. An observatory is not likely to encounter zoning ordinance issues, aside from adhering to setback distances from roads. The permission to build any structure is called a *building permit*, and to get it, one submits an application to the code official of the town or city. With the application, one submits plans of the structure, and a plan of the plot, showing how the structure will fit in. The code official will check that the plans are in line with the state building codes, and, after construction, a building inspector will come and fill out a report to verify that the structure has indeed been built "according to code".

Wherever you live, it is important not to be put off by the planning procedure or building codes. The cost of gaining permission is very low compared to the material costs of any observatory project, and in most cases the procedure will be a formality. There will probably be no objections in principle to what you intend. You will also probably find that the local officials have never had to deal with an application to build an observatory before, and will be slightly bemused by the whole thing. The work of the officials is intended to help you to build a safe structure, which will not prove a risk to anybody else, particularly should it go out of your hands, being passed to a subsequent owner of the land. As an aside, since most observatory owners will not wish to give up their observatory if they move house, but to take it with them, assuming it has been constructed in such a way as to make that possible, it is important to stress that it must be specified, at all stages of a house-selling transaction, that the observatory is not included in the sale. If it is not specifically excluded, it will be assumed to be included, and more than one astronomer has thus accidentally sold his observatory.

So much for the generalities of observatory planning. We must next consider the vast scope of possible designs, and try to analyse their advantages and disadvantages.

CHAPTER THREE

Types of Observatories

A Fully-Portable Setup

Our definition of an observatory is any fixed site from which observations are made. Hence we should first consider the case where there are no permanent fixtures. Even if the mounting and telescope remains portable, there are certain measures which can be taken to make observing more convenient and productive, if it is regularly carried out from a fixed spot.

One of these is to have some kind of mark or marks on the ground indicating how the mount it is to be set up for correct polar alignment, avoiding having to go through a time-consuming polar alignment procedure every night. With a lawn, small stones can be set into the turf with painted marks on them. This will prove quite accurate. Alternatively, a small concrete area could be laid to provide extra stability to the tripod, and depressions for the tripod feet could be made at the casting stage. With most tripods and mounts, these need only be approximately accurate – fine tuning of the polar azimuth alignment can later be made on the mounting itself and then locked in.

Another simple convenience is to have all ancillary equipment gathered together in a briefcase or chest, so that it can all be taken out in one go, or with not too many journeys from house to site. It would be a nice idea to make this box or chest convertible into a small table for notebooks, eyepieces, cameras etc. Otherwise, a lightweight table and observing chair can be conveniently stored in a shed and taken out each time (Fig. 3.1). Observing from a lawn, as here, is more comfortable than observing on concrete, as heat is not transmitted away

Figure 3.1. Basic minimum observatory, with a 20 cm (8 in.) SCT set up on a lawn, with plastic table and chair, and rigid dew-shield, not fitted.

from the feet so fast, but good, damp-proof footwear is a must. Lawns produce good local seeing, as grass emits little infra-red radiation. However, even the best tripod will not be very stable on a lawn. In my opinion, a tubular steel tripod, such as the common type shown here, is better than a wooden one. It is stiffer and more easily adjustable for height and uneven ground.

In recent times, observation via video or other camera from a warm indoor room has become understandably popular. If a fixed observing site is used, permanent wiring can be put in, with a wall or floor-mounted connection box, into which the camera can be plugged. It is then necessary also to have extension control connections to the telescope motors, so the telescope can be driven from inside, and, probably, electric focusing, also operable from inside, and possibly other functions, such as motorised filter-changing, similarly arranged. All this can be combined into one unit, with basic electrical skills. This sort of arrangement is similar to the type of observatory and control-room setup discussed in Chapter 6.

The big disadvantage of the fully-portable setup is the time taken to set-up each time, and the discouraging labour of carrying the components of the telescope to and fro. This will certainly result in many observing opportunities, which would have been taken from a built observatory, being missed. Any telescope larger than 15 cm (6 in.) aperture on a German mounting is almost certainly going to require breaking down into optical tube, mounting, counterweights and tripod before moving, then screwing back together. A fork-mounted telescope of the same size will be more portable, because of the lack of counterweights. Then there is the carrying about of furniture, power supplies or batteries, and computers, if used for imaging or other purposes, to be considered. A Dobsonian telescope is the easiest of all to use portably, but is more or less limited to low-power visual observation.

On the other hand, if the observing site is small and surrounded by obstructions such a tall trees, which cannot be mitigated, then a fully-portable setup might be the best option for maximising observing opportunities. Small, short focal length telescopes, particularly refractors, combined with modern imaging and computing technology can produce remarkable results, particularly in the area of wide-field deep-sky imaging (Fig. 3.2). If this is your area of interest (which you will be able to exploit most fully only if you live in a dark location), then the fully-portable setup may prove practical.

Many keen observers, with only temporary access to the observing site, or living in flats, or observing from a balcony or flat roof, will have no option other than a portable set-up. It may be worth mentioning here the many remarkable lunar and planetary images that have been taken with telescopes set upon balconies, by observers living in densely crowded cities, mostly in the Far East (Fig. 3.3). Such circumstances are always non-ideal however, and good results, as already mentioned, cannot be obtained from above, or near to, heated buildings in cold countries.

A telescope stored inside a heated building will require a cooling-down period after it has been taken outside before it is usable, particularly in the winter. The length of this period depends mainly on the mass of the objective, but also on other details of the telescope design. Fans built into the tube or mirror cell can help reduce the cool-down time. Storage in a clean, dry garage or shed is usually better than indoor storage, however.

Some observers choose to leave a tripod and mounting in the garden or backyard all the time, and only carry the optical tube in and out. The mounting

Figure 3.2. M42 taken by Dave Grennan using a Celestron Onyx 8 cm refractor and Atik 16 HR CCD camera – a good example of the type of wide-field image at which such small, portable refractors excel.

Figure 3.3. This highly-detailed image of Mars was obtained by Chris Go using a Celestron 11 28 cm SCT, a FireWire webcam, and a laptop, from his balcony in a high-rise block in Cebu City, Philippines.

may be covered to protect it from the rain, but most possible coverings tend to trap moisture and promote corrosion, and they are also prone to being blown away. This is a less permanent version of our next category of observatory.

Removable Telescopes with a Fixed Mounting

A halfway-house to a roofed observatory is to have a permanently-fixed telescope pier and mounting outside, and a removable telescope, stored inside or in an outbuilding. Some people leave most of the telescope outside, removing the optics indoors after every session. This does not sound very convenient to me – it would require complete re-collimation every session, for a reflector.

However, there are distinct advantages to setting-up a permanent metal or concrete pier to carry the mounting, on proper foundations dug into the ground, and creating around that a comfortable observing surface, of which the best kind is a wooden deck. This might be regarded as the simplest and most economical type of observatory. The labour of carrying around the heaviest parts of the telescope, the mounting and counterweights, is removed, and there is no necessity for repeated polar alignment or levelling of the mount.

Observing on damp grass is no fun. A telescope is not completely stable on a soft surface, one's feet can get wet, small objects (set-screws holding eyepiece barrels or finders are particular culprits) fall off and get lost, perhaps for ever, and one can't kneel down. On the other hand, a concrete or slab surface is also poor. It will generally transmit too much vibration from roads, it will be very cold to stand on (concrete rapidly conducts heat away from the observer), and, worst of all, optical equipment dropped on it is not likely to fare well. If you are anything like me, you will drop eyepieces, prisms, filters and cameras with sad regularity.

A wooden deck supported on joists forms a good observing surface. Decking timbers are universally available from DIY outlets, and the construction involves only rudimentary carpentry skills of sawing and screwing. At this stage, I would suggest that anyone considering building an observatory of any kind, who does not have such skills and experience, and would like to acquire them, goes and looks at some good books on woodworking and other DIY topics, which abound in all bookstores, and possibly enrols for a course at their local adult education centre, if available. The knowledge acquired will certainly prove very useful in the long-term, even if they eventually decide to buy an observatory, or have one made for them. More will be said about construction methods in Chapter 5.

If it is on grass, the deck will need to be raised above the ground by bricks or concrete blocks to allow for ventilation. If on concrete or slabs already, the joists will be sufficient. Both proper decking and joist timber is pressure-treated and already contains preservative to protect it from rot, but this will only be effective for a short time. It must be treated with a wood preservative shortly after construction, and annually thereafter.

More will be said about telescope piers and foundations later. The point to note for the moment is that, with the possibility of leaving an equatorial mounting set up permanently, the polar alignment can be made perfect, and this is invaluable for all serious astronomical work, in imaging, photometry, astrometry and surveys. It also makes faint objects much easier to find, whether using traditional setting-circles, digital ones, or a GOTO system. Of course, this type of observatory requires that the telescope is easily removable from the mounting. This will generally be true with small to medium-sized telescopes on German equatorial mounts, but not necessarily true with fork mounts or other types. Of course, with large telescopes, carrying even the optical tube about becomes too difficult.

If the tube is stored in a heated house, the remarks made before about allowing it to thermally equilibrate with the outside air before observing apply. However, the telescope will be protected from the effects of damp, corrosion, insects and arachnids (spiders), which plague most observatory telescopes. The aluminium coatings on reflector mirrors will probably last much longer, and, of course, the most valuable parts of the telescope will be safer from theft. One well-known and highly-respected English observer kept his valuable refractor outside, but took the optics in after every session. Very sadly, the brass tube was stolen, and never recovered.

With the fixed mounting approach, the problem remains of how to protect the mounting from corrosion and other damage when it is kept outside. Flexible "scope-covers" are sold by some astronomical equipment suppliers, and these could be tied around the pier with rope, but I haven't tried them. A plastic car-cover is a possibility I have tried, but I found this tended to trap the dampness that accumulated during observing sessions, and promoted rust. A removable wooden box might be the best solution, purpose-built to fit over the mounting on top of the pier.

Such a semi-observatory can be supplied with mains electrical power, subject to the usual safeguards for outdoor electrical installations (contact a qualified electrician, and see the advice in Chapter 8 of this book), eliminating a major disadvantage of portable setups, the requirement for batteries or inconvenient, and possibly unsafe, mains extension cables. The problem with using batteries for astronomical equipment is that they will always fail at the crucial moment, just when conditions are perfect for obtaining that sensational observation or image. Even if regularly re-charged, they can lose power dramatically in low temperatures. Moreover, keeping them charged-up is a chore. Temporary mains extensions from another building are not to be recommended, as it will probably be necessary to split the power between different devices, for example, a telescope mount and a laptop computer, with a non-waterproof adaptor, that becomes exposed to damp. Such an arrangement poses a significant risk of fatal shock to the observer. Proper waterproof outdoor sockets, installed as part of an observatory project, additionally often become useful for operating gardening equipment.

The disadvantage with the semi-observatory is that it will, like the fully-portable setup, still most likely require a lot of journeys to be made to carry equipment in and out of the house or other storage shed, thus losing most of the setup time-saving advantages of a proper observatory. Some of these items are: eyepieces, diagonals, filters, cameras, computer, an observing chair, and observing table, power adapters, cables, and dew heaters. The list really goes on and on. The more complicated your astronomy becomes, the more inconvenient becomes the rebuilding of the system each session. Having all these things ready to hand, more or less in place, in a roofed building, is one of the main joys of owning a full observatory. Of course, some of them can be gathered together in a briefcase, but some cannot.

Run-Off Sheds

Run-off sheds are the next stage of sophistication of the amateur observatory. With the run–off shed, sometimes called a roll-off observatory, the telescope and its mounting are permanently assembled, but used in the open. A movable covering protects the equipment when not in use. This covering usually takes the form of a small shed-like structure, either in one or two sections, that moves on wheels constrained by rails or tracks.

The run-off shed preserves the completely open character of observing with a portable telescope. The whole sky can be seen (or as much of it as the local horizon allows), and many observers value this highly, and would not swap it for the environment of a dome, where only a narrow slit of sky is visible at any instant, despite the alluring comforts of the walled and roofed observatory. If a fireball streaks across the sky, making an unexpected display, or a grand aurora occurs, or a bright artificial satellite passes over, the observer in the dome may well miss all these unpredicted events, but they may be seen by the observer with the run-off shed. The run-off shed itself affords no protection from the wind, though windbreaks can be separately constructed, and it doesn't shield from local artificial light sources.

Equipment in a run-off shed suffers badly from the dew in damp climates, more so than portable equipment, which dries off after it is taken in. Run-off sheds tend to trap the dampness that accumulates during observing sessions, retaining it into the following day, unless considerable attention is given to ventilation. They suffer particularly from this as their volume tends to be quite small. Another factor contributing to the damp problem is that rainwater flows underneath the shed. I have found more problems with electronic equipment kept in a run-off shed than when housed otherwise, and perhaps the run-off shed is a less appropriate solution in these days of high-tech observing setups than it was in simpler times.

A problem with using a run-off for a "high-tech" set-up is that the electricity probably will have to be routed via the pier, and thus all electrical equipment will end up clustered round the pier. This can be a bit messy and "in the way" (Fig. 3.4). A fixed-wall observatory is more flexible with regard to wiring.

However, a run-off shed represents a big step-up in convenience from a portable telescope or fixed mounting arrangement, and is a type of true observatory. Run-offs can be made quite sophisticated. A deck can be constructed, similar to that described in the previous section, as a working surface (Fig. 3.4), and shelves and a desk can be installed in the moving shed. A desk in the shed would be useful for webcam imaging of the Sun or planets in daylight using a laptop, the shed shielding the computer display from the daylight glare. Items such as an observing chair can be stored in the shed and brought out for use.

There are several possible patterns of run-off shed (Fig. 3.5). A one-piece design is possible (A), with a hinged door or doors (two doors will save space). My run-off shed is to this pattern, and a complete description of this will be given in Chapter 9. The one-piece shed has doors that are opened; then the whole building, or box, is rolled clear. Mine is arranged so the doors can then be bolted in place so as not to flap about. This design has the disadvantage that it can be heavy to move. An advantage is that it has no roof joins which could leak. An alternative to using hinged doors is to have a removable side (B), but then one has to lift this clear to start with, and put it somewhere else, which also might be strenuous. Patrick Moore's famous run-off is of the two-piece variety (C). The sections of this design will be lighter than the one-piece, and hence easier to move, but there may be difficulty in waterproofing the join.

Run-off sheds generally run off on small wheels, which could be metal or plastic (but they cannot be swivel castors, which, undesirably for this application, spin on their axes). These wheels generally run on metal rails made from angle-iron. This is not all that easily available in the required lengths, and will need to be ordered from a metal fabrications supplier. The length needs to be at least twice the side-length of the shed, and probably a bit more for adequate working-room. A straight rail is, however, much easier to obtain than the curved rail required for a rotating dome. Martin Mobberley has used plastic rails successfully in one of his observatory designs, of which more anon. Another alternative to a metal rail might be a channel in concrete, cast *in situ* using wooden formers, which are removed when the concrete has set. I have not seen this put into practice. However, I came up with a design which avoids a rail or channel entirely, instead

Figure 3.4. With an observatory without fixed walls, wiring and electronics tend to get crowded round the pier.

running the shed on a constrained track on wooden decking. This works quite well. Martin Mobberley has exploited the variant of moving the telescope on rails, rather than the shed, and has successfully used plastic rails for this purpose. These examples will be described in more detail in Chapter 9. The most important points about rails are that they must be set, and remain, perfectly parallel, level and stable, but metal needs to be allowed scope to expand and contract with the changing temperature.

A run-off shed is not a very complicated construction, and it is a project that most people with some practical know-how could tackle. Alternatively, a carpenter or builder could easily construct it, though they would have to be well-briefed as to its operation and purposes. Run-off sheds are not made commercially as a regular product, so far as I am aware. The constructional problem of the run-off shed is that it has to be light enough to move by hand, but rigid enough to withstand the

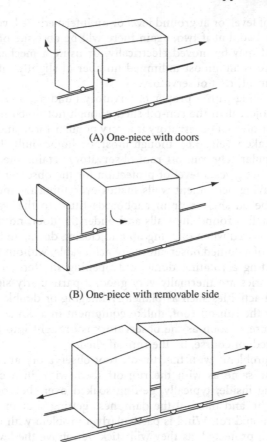

(A) One-piece with doors

(B) One-piece with removable side

(C) Two-piece

Figure 3.5. Possible types of run-off shed.

force of moving it without distorting, and that it cannot be braced by a floor as can a normal shed (though the rails brace it slightly). It is further weakened by lacking one wall. If timber, therefore, it requires substantial internal bracing, which needs to be arranged not to restrict clearances too much. Probably the best construction for a run-off shed is a welded steel framework clad in plastic or wood. Olly Penrice has constructed two such sheds at his astronomy holiday centre, Les Granges, high in the mountains of Provence, France, starting with no knowledge of welding whatsoever. Most DIYers are happier working in wood, however, and most run-offs are timber, roofed with some waterproof material.

Run-Off Roof Observatories

Run-off (or roll-off) roof observatories are the next stage in sophistication of the amateur observatory. The walls are fixed (or largely fixed), and the roof moves, probably in the horizontal plane, using wheels running on rails. The rails may be just below the roof level, or at ground level or an intermediate level. The roof can be in one part or two, but if two, again there will be an issue of waterproofing the join. The roof may be moved electrically, or using a mechanical winch, or just pushed. There is an almost unlimited number of slightly different possible variants of the run-off roof observatory.

Here we are dealing with a proper observatory building, and a rather bigger constructional project than the run-off shed, though not much more difficult in principle. The run-off roof observatory is a very popular form, and can be bought from specialist makers (Fig. 3.6), though most are home-built. They are always square or rectangular. The run-off roof observatory retains the all-sky view of the run-off shed, but gives a level of protection for the observer from wind and external light. Having permanent walls makes everything far more convenient. Accessories can be on shelves or in cupboards fitted to the walls, permanent wiring can be installed round the walls and under the floor, nothing apart from the roof need be moved when opening up and closing down, and, in essence, all the conveniences of a domed observatory are achievable, without the engineering challenge of creating a rotating dome and opening shutter. In addition, run-off roof observatories are thermally very good, a particularly significant factor for seeing-critical activities such as planetary imaging or double-star observing. Everything under the run-off roof, unlike equipment in a dome, rapidly falls to outside temperature as soon as the observatory is brought into operation. This advantage is shared, of course, by the run-off shed.

The principal problems with the run-off roof observatory are dew and wind. The former is the same as with the run-off shed, with, in a climate like the British, everything inside, typically, getting soaked over the course of a long, clear winter's night, and then all the dampness getting shut in during the day to cause corrosion and rot. Wind is particularly a problem with long telescopes, which are not well protected, as they will stick out above the level of the walls, and suffer a turning moment from the wind striking their upper end but not their lower end. Hence the run-off roof can lead to worse wind problems than having the same telescope in the open.

It might be thought that the walls would raise the horizon to a high altitude, and they always do limit the horizon somewhat. The situation must be considered in each local case, and carefully optimised. Very common is the idea of having an upper section of the south wall folding down after the roof is moved off, so allowing a clear southern horizon. Most commonly also, the roof rolls to the north, but this is not always possible. In designing my run-off roof observatory, I decided not to have a folding wall, for the sake of simplicity, speed of operation, and structural strength. Instead, I made the walls low, so that an average-height person cannot stand inside with the roof on, but the telescope can still be stowed away, almost horizontally, and on the meridian. The way the walls were arranged with respect to other local obstructions (trees and buildings) means that very little usable sky is obstructed by the walls, or open roof, in this case. But in any case, it is not really an issue to be worried about too much, as very little useful telescopic observation can be undertaken at an altitude of less than 25°.

For my first observatory, which I built at the age of 17, I created an almost cubic structure that had a two-leafed roof that opened on hinges, like the petals

Figure 3.6. Two run-off roof designs manufactured by Alexanders Observatories of England, a small 2 m (6 ft) square example, and a large 4.8 × 3 m (16 × 10 ft) observatory with a warm room.

of a flower (some historic shots featured in Fig. 3.7). Since this observatory was only 1.5 m (5 ft) square (far too small), the roof sections were not too heavy to be opened in this manner. However, it never worked well. The join proved impossible to waterproof satisfactorily (as can be seen, the rainwater flowed down the slope into it). In any case, I am sure such a design would not be practical if made to a sensible size; the roof would be too difficult to open. Since then I have been rather against two-piece roofs and two-piece run-off sheds. My second observatory was the run-off roof shed I use today, described in Chapter 9, with a one-piece roof. It never leaks at all, unlike most amateur domes.

Figure 3.8 shows some of the possible variants of the run-off roof idea. In (A) we see a pent roof shed which rolls to the south along the line of the apex,

Figure 3.7. My first observatory, a 1.5 m (5 ft) square hinged roof structure, built in 1981 (The board at the back was to screen a streetlight).

taking with it the eaves sections. In (B) we see a design where the roof slides northwards and perpendicular to the apex, and then the top of the south wall folds down. In (C) we see a flatter, pent-roof shed, where the whole upper section of the walls moves with the roof, to leave very low walls. Case (D) is the slightly unusual design of my run-off roof, which moves perpendicular to the slope of the roof, on rails of different heights. The wheels operate on a flat plane, as with the other designs. It is also possible to have a pent roof which slides in its own sloping plane (i.e. not on the flat), if it is operated by a winch mechanism (E). The diagrams show one-piece roofs, but two-piece roofs, sliding in opposite directions, are also possible. However, this is usually not necessary in terms of ease of operation. The roof can be made quite light, so even a large run-off shed

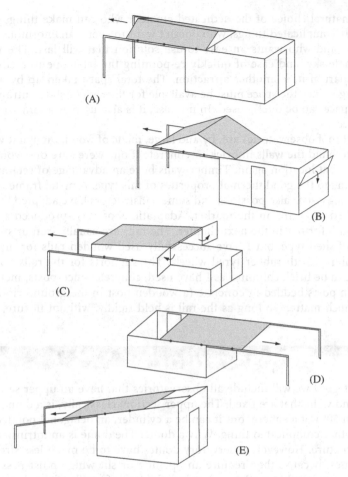

Figure 3.8. Various possible patterns of run-off roof observatory (north is conceived as being to the left).

can normally have its roof moved manually in one piece, if it is well enough made and maintained, thus avoiding a leaky join.

Types (A), (D) and (E) are the strongest constructions, being uninterrupted boxes, while (B) and (C) do a better job of clearing useful horizons. Type (C) is like a run-off shed elevated into the air slightly. This is going to be the hardest type to move manually. The diagrams are by no means exhaustive of the designs that could be, or have been, used for run-off roof observatories. The choice will be made on the basis of local circumstances and the instruments to be used, and there is huge scope for invention.

The popularity of run-off roof observatories is down to their relative simplicity of construction and engineering compared with domes. They supply a proper, fixed observatory without the requirements for fabricating curved components, and solving the very difficult challenge of making a dome slit that opens easily, is large enough, is waterproof, and doesn't fatally weaken the dome. They are thus the natural choice of the dedicated amateur who can make things, but not excessively complicated things, who doesn't want to spend an enormous amount of money, and who wants an easy-to-use solution that will last. The visibility of the whole sky, and ease of quickly re-pointing the telescope to a completely different part of it, is another attraction. The total space taken up by a run-off roof is large, since the space must be available for the roof to slide untrammelled, but this space can be dually-used. In my case, it is also used as a tomato-growing sun terrace.

Run-off roof observatories are, by and large, made of wood, though it would be possible to build the walls in brick or concrete if one were sure one would never be leaving that location again. Timber walls have an advantage of retaining little heat, adding to the good thermal properties of this type. A metal framework clad in something else is also possible, and some constructors have adapted the plastic storage sheds that are on the market. Adaptations of mass-produced sheds will be dealt with further in the next chapter. The rails are usually iron or steel, as in the run-off shed type, but I have successfully used wooden rails for my run-off roof, combined with rubber-tyred wheels. The supports for the rails away from the shed can be brick columns, as I have used, concrete fence-posts, metal posts, or wooden posts bedded in concrete (a wooden post in the ground risks rot), it doesn't much matter, so long as the rail is held rigidly, without flexure, in use.

Domed Observatories

In this category we will include all observatories that have an upper section that rotates, and walls that are fixed. The upper section, classically, is a dome, slightly more than 50% of a sphere, but it can be a cylinder, an octagonal pointed shape, or some other complicated thing. Why a dome? The dome is an intrinsically very strong structure. However, observatory domes have to be much less strong than other domes, because they require an opening or slit which must pass beyond the zenith. Taking this section out causes the loss of much of the dome shape's architectural strength. A uniformly curved dome should shed water uniformly in all directions. It should shed running water away from the opening. (However, many domes have a problem with rain blowing in under the shutter.)

Domes largely or partially solve the problem of condensation that afflicts the simpler observatory types. Condensation occurs because surfaces, particularly metal and glass ones, radiate heat away at to space at night to the point at which they become colder than the surrounding moist air. They then push the air in contact with them to below the dew point, and water condenses on them. With domed observatories, the dew normally forms on the outside of the dome, leaving equipment inside dry. However, this does depend on how close to the dew point the air is already, how wide the dome slit is, and on how long the observing session lasts. The dome may only delay condensation inside. In general, a wide slit is desirable, as this means less frequent dome rotation (unless the dome is automatically driven), and the wider the slit, the more the dome behaves like a run-off roof observatory.

Domes protect the telescope from vibration by the wind, and, in keeping wind off the observer, make for a much more comfortable observing experience, even though the temperature in a dome must be allowed, and indeed forced, by fans and ventilation if necessary, to reach the outside temperature. The biggest observing disadvantage of domes is their tendency (particularly metal ones) to retain heat, or to funnel warm air out of the slit, when they are first opened up, in the manner of a chimney, severely impairing the seeing at the telescope. A dome must be opened-up well before observing, particularly after a warm day. Run-off sheds and run-off roof observatories allow the equipment and its environment to cool down much faster.

Domes are manufactured commercially in galvanised steel, aluminium alloy, and fibreglass. Fibreglass is the lightest and cheapest of these. Lightness is a quality to aim for in a dome, as it is a moving part, like the roof of the run-off shed, whether it is mechanically, electrically, or manually turned. However, this leads on to the fact that domes can, and do, blow away. The pressure of a Force 10 wind on a fairly small dome (4 m or 13 ft diameter) is of the order of a tonne (1000 kg or 2200 lb.). Domes need both to be strong, and strongly-retained by their base-ring. The base-ring is normally a circular rail of steel, welded together. This is a component that has to be made specially, and can be difficult for the amateur dome-builder to obtain. Dave Tyler (Chapter 9) used a base ring made out of cut sections of aluminium sheet instead.

Galvanised steel domes are fairly cheap and very strong, but heavy, and tend to rust where the material is drilled for fixings. Aluminium alloy (of which the most useful type is known as duralumin, used extensively in the aircraft industry) is really the ideal material for a dome, being light, strong, easily worked-with, and corrosion-free. However, there are few amateur examples of its use. The well-known UK observer Maurice Gavin clad his wooden dome roofs with aluminium plates, which made an attractive and durable effect.[1]

Many ingenious methods of fabricating domes, or approximately dome-shaped structures, from wood and metal and fibreglass, have been devised by amateurs and published in the previous reference and on the internet (see Appendix 2). There is no point reproducing this material here, as it is so easily available. It needs to be said that flat sheet materials cannot really be made into true dome sections, as that would involve bending them in two planes at once. One class of solution to this problem involves cutting "gores" – narrow, triangular-shaped sections of plywood or metal, that can be bent to be fitted over a wooden or steel frame, their points meeting at the top. Another class of solution involves simulating a dome-shape using flat sheets of board or metal on a frame. A fibre-glass dome is normally cast in sections, which, again, are commonly attached to a frame. This type of dome can be truly spherical, if the moulds for the fibre-glass sections can be given the right curves. However, there are few advantages to having a truly spherical dome. Some constructors do not aim for a dome-shape at all, but for the simpler "pepper-pot". Patrick Moore's famous dome covering his 38 cm (15 in.) reflector is an example. It was made by his local blacksmith, but sadly there are not many of those about these days. The roof of this pepper-pot dome is a shallow cone-shape, providing the necessary drainage

---

[1]See *More Small Astronomical Observatories,* Patrick Moore (ed), Springer 2002, on the CD ROM of the first book.

slope. Polygonal designs with numerous flat faces can be another outcome of dedicated woodworkers' efforts to approximate the dome shape in inflexible sheet material.

The wheels on which the dome rotates usually are arranged to retain it in two directions – in the horizontal plane, and also to prevent it from lifting off. The wheels may be fixed to the base ring of the dome, with the rail fixed on top of the lower section (the dome wall), or the arrangement may be the other way round, with the wheels attached to the dome wall, sticking upward, and the rail attached to, or integrated with, the dome base ring. Most domes are rotated by hand, pulled with ropes or handles, but a few have, like Patrick's, a toothed base ring that is engaged by a cog wheel, turned with a crank. Such a mechanism can be motorised; however, a toothed base ring is extremely time-consuming to make. Motorisation is offered as an option on some commercial fibreglass domes as well, which use rubber rollers to move the dome by friction. It is possible to even have a dome computer-controlled (as is the norm in modern professional observatories), so that the slit is always aligned with the computer-driven telescope. Such systems can be purchased, but some amateurs have also designed and built them for themselves. (Many amateurs who have set up computer-controlled imaging systems, however, such as the record-breaking British supernova hunter Tom Boles, have preferred the simplicity of a run-off roof observatory.)

The slit or opening in a dome needs in any case to be fairly wide, at least twice the aperture of the telescope, and probably more with a manually-pushed dome, to prevent the dome shadowing part of the aperture and to avoid it having to be moved too often. Astronomers who are not used to working in a dome generally find it, at first, very disorientating. So little of the sky can be seen at one time, it is difficult to find one's way about the sky by the normal outdoor methods of recognising familiar constellations and asterisms. As the dome is rotated, one can lose one's sense of direction. These problems do not apply, or course, to a fully automated, computerised system. But that presents a very different experience to the traditional amateur astronomers' methods of working and relating to the sky above his head – one that many are not looking for (and of course it is expensive).

The construction and waterproofing of the slit shutter is, as has been indicated, a significant problem. Some slits open by sliding a curved shutter up and over the adjacent dome on semi-circular track sections. This looks professional, but is difficult to make. Others slide the shutter to the side on a straight track, and others arrangements use a flap opening on hinges. The slit-closure can be in one part or two. If it opens on hinges, it may be necessary to leave it waving about in the air like a sail, unless the range of the hinges is engineered to be greater than 180°. Then, getting it back home can be a problem. Shutters in higher observatories are often opened using rope and pulley arrangements, or using poles with hooks on. Polygonal and drum-shaped dome designs make the slit-cover problem more complicated, as the cover has to be in sections, perhaps opening in different ways. With a drum there must be a section of the vertical face that opens, as well as a section of the flat, or sloping, roof. "Up and over" shutters have a possible advantage over sideways-opening shutters in that they do not need opening completely to view at low altitude, but it is not that often that this advantage is likely to be utilised.

The lower section of a domed observatory, the dome wall, is usually quite low, often about 1 m (3 ft) high, with a door that can only be entered by the observer bending double to get under the rail. This is to keep the telescope fairly low, so ladders are not required for observing, while allowing it to point almost horizontally out of the slit. This is less true of domes for some telescopes, such as classical refractors and long-focus Cassegrains, which are generally mounted quite high, so the dome wall can be higher. The dome wall can be of a completely different construction to the dome itself; it can be wood, fibreglass, metal, or masonry, if you have it in mind never to move from that location again. Many people build brick or breeze-block dome walls themselves and then buy the rail and dome from a manufacturer to place on top. This saves buying the manufacturer's dome wall, and is obviously stronger, and completely permanent.

The main issue with commercial domes, as mentioned before, is that most of them are just too small. If they are large enough (3 m or 10 ft diameter at least), they are very pricey. Some manufacturers have attempted to make their domes expandable with optional side compartments, for computers or for storage. These jut out from the walls, not reaching the ground, in the manner of panniers. They can be added on after the observatory has been set up, as needs increase and funds allow.

Another way forward with a commercial dome might be to build your own dome wall of say 3.3 m (11 ft) diameter, which could then be flat-roofed with a annulus of material, and combined with the dome and track of a smaller manufactured observatory, thus giving adequate space at a reasonable cost, with a tried and tested dome design.

A few domed observatories have been built which do away with the difficult-to-make perfectly circular metal rail by having the wheels on the dome base run on a smooth finish on top of a masonry wall, using something like ceramic tiles. I do not know how successful these have been.

The lower section of a domed observatory need not be circular. It can also be octagonal, hexagonal or square. These shapes are obviously much easier to achieve in wood. Making the lower section square, with side length equal to the dome diameter, allows much more space inside and, since most furniture is not made with circular curves, is much more convenient when installing a desk.

Totally-Rotating Observatories

It has been noted that the usual amateur dome proportions result in a very low and awkward door in the fixed wall. A way round this is to have the whole observatory rotate in one piece. The floor may be fixed and the observatory walls may rotate on a rail, or the floor may rotate as well, pivoting about the central telescope pier, which is, as in all good observatory designs, kept completely separate from the observing floor. This is probably the most disorientating design of all: at least in the standard dome, the position of the door is fixed.

The totally-rotating idea allows the building to be a simple box shape, perhaps with a pent roof, or, at least, some shape much simpler to make than a dome. It also allows a strong construction, as the walls, floor and roof can be mutually bracing. The Achilles heel of this design tends to be the roof flap, which, being a section of a flat roof surface, is difficult to make waterproof. The building can also be hexagonal or octagonal, though adapting the roof-shapes of these buildings to form a suitable slit going beyond the zenith is not easy. Factory-made garden conservatories have been adapted, though it should also be mentioned that having a lot of glazing in an observatory is distinctly not a good idea, due to the well-named "greenhouse effect".

Novel Designs

While the foregoing essentially covers all the working observatories I have come across or heard about, is always possible that fresh thinking will produce usable new designs, especially with the continual advance of materials technology.

At the time of writing, a new type of prefabricated observatory has recently been announced by its manufacturer, SkyShed of Canada, but is not yet fully tested. Cleverly called the SkyShed POD (Personal Observatory Dome), it is a

Figure 3.9. The SkyShed POD (a prototype version).

domed observatory made entirely from high-density plastic sections that can be easily self-assembled (Fig. 3.9). The novel feature is the opening, which is something like an eyelid, opening one entire half of the dome to the sky. The upper section is also rotatable. The observatory is designed to be portable – easily dismantled and reassembled "in the field" for remote, dark sky observing expeditions. Because of this, no doubt, it is on the too-small side, at 2.26 m (7 ft 5 in.) internally, but this can be mitigated to some extent by adding storage compartments, external to the basic walls, in a modular fashion. The idea is very interesting, though I suspect the all-plastic construction might lead to ventilation difficulties when closed. Moreover, the 50% dome opening will result in no protection from dew (it will be like a run-off roof observatory from this point of view), and it is really too small to be regarded as a serious observatory, except for automated imaging purposes.

To try to be fair on this, the kind of circumstance in which the POD and other small domes might work tolerably well for normal observation is shown in Fig. 3.10. Here, the POD is on a patio deck, right up against a house. There is only a view in the direction away from the house (which had better be south in the northern hemisphere), so the telescope is placed up against the south side of the observatory, allowing plenty of room on the viewing side. Things are made more practical by the fork mounting, which takes up much less space then a German mounting would do. Only observation around the meridian is comfortable. Of course, if you moved the telescope around in the observatory, you could access the whole sky: but, for an equatorially-mounted instrument,

Figure 3.10. The prototype POD in use on a deck by a house.

Figure 3.11. The OWL (Over Whelmingly Large) telescope concept of the European Southern Observatory (*courtesy ESO*).

this defeats a principal purpose of having the observatory. The POD is perhaps an accessory more comparable to the observing tents mentioned earlier than to regular observatories.

A different fold-back dome design is offered by another US company, Astro Haven. Their "clamshell" fibreglass observatory design is quite long-established. Here, the dome does not rotate; both halves of it fold back on a common hinge, like a two-sided eyelid. The clamshell is made in sizes from 2.1 m (7 ft) to 6 m (20 ft), and also in non-circular, elongated designs. The moving sections can be hand-operated, or motorised. The roof may be retracted completely, to give all-sky visibility, or partially, to provide more protection from wind or artificial lights (though there will be little protection from dew). The design allows a lower horizon all round to be attained than is possible with a run-off roof observatory, while still preserving the benefits of having fixed walls.

There has been an interesting tendency recently for the most advanced professional "mega-observatories" to come to resemble favourite amateur designs writ large. The engineers of these structures have been given second thoughts by the expense and difficulty of building spherical domes of the required sizes, and have, for example, made the buildings that house the four 12 m reflectors of the European Southern Observatory in the Chilean mountains somewhat resemble Patrick Moore's blacksmith's "pepper-pot". And the concept design for a future European 100 m "Overwhelmingly Large Telescope" (this name seems to be intended as something of a joke) is in the nature of a huge run-off shed arrangement (Fig. 3.11). Be this as it may, we need now to get down to the details of how to make a reality of your own observatory plan.

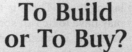

CHAPTER FOUR

To Build or To Buy?

A decision will need to be taken on whether to build your own observatory from scratch, to adapt an existing building, to buy a commercially-made building and adapt that, to have an observatory built to your design, or to buy one "off-the shelf". Cost considerations will be significant, but so will other factors. Some people do not have the time and inclination to make things themselves or to learn new skills. Others get great enjoyment and satisfaction from tackling practical problems and finding a way through them through their own ingenuity. It is certainly always advantageous, if you do not possess useful skills yourself already, to cultivate your relationship with family, friends and acquaintances who do have such skills, and perhaps use them professionally. Knowledgeable people will usually be only too keen to advise and help, when presented with an unusual challenge. Architects do not design observatories every day, neither do builders build them every day. However, this fact also means that you will have to be very thorough in making these non-astronomers understand exactly what is required.

Costs of DIY and Ready-Made Observatories

It is as well to be realistic. The costs of large projects over-run, and the costs of small projects over-run, almost always. That is the way of the world. However, an amateur observatory generally does not have to be completed to a strict timetable. One can carry on observing productively with portable equipment in the interim, and an observatory can be used in an unfinished state. It is usually hard to say when an observatory is completely finished, and many never are. There are always improvements that can be made with experience.

If buying an observatory ready-made, the cost may seem clear, but in reality many expenses that are not included in the basic delivery of an observatory structure will have to be taken into account. There is concrete for foundations and the cost of labour or help laying them. There is the cost of the telescope pier and of electrical wiring, which may have to be run some distance from the nearest supply, and associated components. It is much harder to calculate the cost of a home-built observatory. Apart from the materials, many tools will probably need to be bought if similar work has not been attempted before. My experience has been that one cannot say with any confidence that constructing your own observatory is going to be cheaper than buying one, contrary to what most people will say. It will almost certainly be cheaper than a commercial fibreglass dome, which will be at least US$8000 or £5000, at 2007 prices, for an adequately large one. Commercially-made, purpose-built, run-off roof sheds of adequate size can be bought for about US$6000 or £3500, if you have all the installation work done, or $4000 or £2500 if you do much of the assembly yourself. None of these prices include the "extras". (See Appendix 1 for a list of observatory manufacturers.)

If we want to compare these prices with DIY costs, it is not easy. There is, I have noticed, a tendency for home observatory constructors to understate the costs of their projects enormously – as if it were a badge of pride to say, "I knocked this up for less than £100", or whatever. This has been referred to, in Britain, as an aspect of the "post-war austerity" mentality: that people who will willingly spend thousands of pounds on, say, a car, will often claim they are spending almost nothing on their hobby. For example, when I looked at the details of an observatory project that was published in an astronomy magazine recently, it was immediately obvious that the claimed cost of £500 did not include the cost of many vital components.

I would say that, for a costing of a completely home-built observatory, from stock materials, on a basis to compare with the prices above, taking all tools and fixings into account, one would be looking at a figure in the region of US$4000 to US$6000, or £2500 to £3500. These figures are very vague; it is impossible to be precise, as there is so much possible variation in design and materials. But I think they are realistic. I think that for the extras of foundations, telescope pier and electrics, a figure of around US$600 or £400 is likely to be "in the ballpark". I can see many people saying these costings are far too high, but I think they are realistic. An observatory project cannot be truly costed until it has been in operation for several years, and all the problems and "bugs" that arise have been dealt with.

Most enthusiasts will not cost their time. There is the satisfaction, and sheer fun, to be gained out of such a project, which cannot be measured. In addition, by designing and building your own observatory, you will gain valuable skills and knowledge that can be applied to other projects, and you will get a bespoke creation that is right for you – correct for your location and all its peculiarities, and for your instruments (which will, however, probably change, so best not to tailor to them too closely) and observing predilections (which may also change).

You may well get "the bug" and go on to build more observatories, until you run out of space, or goodwill from family and neighbours!

Adapting Commercially-Made Outbuildings

It is a striking fact that the cost of mass-produced sheds and summer houses is far, far lower than the cost of buying the materials to make them – at least 50% less. These are the economics of scale. It is therefore an attractive proposition, in both financial and labour-saving terms, to create an observatory by buying, and then adapting, one of these structures. Square and rectangular pent and apex roof sheds have been adapted by amateurs into run-off roof sheds, and octagonal summer houses have been adapted into domes. It is not proposed to give detailed plans for these conversions here, but many accounts of such work will be found through the web links given in Appendix 2. These will likely only be of general help, as, most probably, you will never be able to find exactly the same shed design that someone else has converted.

Some general considerations will be highlighted here. Timber sheds that are offered for sale by DIY chains are generally very lightly, in fact, rather inadequately, constructed. The main structural timbers are held together, generally, by butt-nailing, which is the quickest and poorest method of timber construction. Such sheds only have strength because of the mutual bracing of the four walls, roof and floor, and they will not stand up long without any one of these. It will certainly be necessary to strengthen them with additional timber or metal bracing, as part of the adaptation. It will be necessary to cut the floor to make a hole for the telescope pier, so this will need reinforcing. The roof can generally be removed quite easily, but will the walls be strong enough to support rails? Probably not, in which case, the rail supports must be made independent. All this is such a lot of work, it might be concluded that it is just as well to start from scratch, and make something designed a priori to do the job well. A construction from scratch is likely to be far stronger and more durable, and the dimensions will be exactly right. Most commercial sheds are rectangular rather than square, which is a disadvantage. However, an adaptation is still likely to be quicker and cheaper.

Adapting a shed is likely to result in a run-off roof that is quite high, and will not allow the telescope to point close to the horizon. The same is true of adapting a garage. It would probably be very difficult to incorporate the popular folding south wall idea in such an observatory. However, this level of obstruction may be desirable if there are many artificial lights around.

The walls of commercial sheds are generally very thin, and have a utilitarian finish inside, being just the inside of the external cladding. This could be treated, by lining the shed internally with hardboard or thin plywood. The wooden sheds manufactured by specialist suppliers are generally much better-made (and more expensive) than those sold in the main retail chains, and less remedial work would have to be done if basing an observatory on one of these. An observatory does not need many windows, and too many, as in summer house designs, will constitute a greenhouse, in which the telescope will get baked in the summer. Windows can be covered, however.

The conversion of an octagonal summer house to a rotating observatory with a slit is a complicated proposition. The proposition is attractive, however, because the result will be far more aesthetically pleasing to non-astronomers than a dome. Waterproofing the roof where it is cut for the slit is likely to be the big problem.

Recently, astronomers have become interested in adapting the plastic storage sheds sold by the main chains. These have the advantages of being low-maintenance, very cheap, and quite strong. A good example has been published in *Sky at Night Magazine*,[1] and another is given in Chapter 9. How well these fare in the long term remains to be seen. Many plastics are degraded and become embrittled by ultraviolet light, in other words, daylight, in the long term. A plastic roof for a run-off roof observatory is a pretty good idea, as it will be light, and in the long term could be replaced if it cracks. This is a long way short of having to replace the whole observatory.

Employing Others to Build Your Observatory

If you do not have the inclination to build or adapt your observatory yourself, and feel that none of the commercially-available solutions are quite what you need, then you may wish to design your own observatory, and have others build it for you, or you may wish also to get help in designing it from an expert.

The way to start would be as if you were going to build it yourself, by drawing out a plan and set of elevations as accurately as possible, with fully-quoted dimensions. The drawback with employing architects, builders and carpenters, is that they will almost certainly not be astronomers or experts on observatories, and will have little understanding of what you are trying to achieve, without a great deal of detailed explanation, in which process vital points can easily get overlooked.

A case of what can go wrong in this kind of process is illustrated by Bill Arnett's (in some ways very impressive and remarkable) observatory "Ptolemy's Café" in California.[2] Bill worked in collaboration with a structural engineer to design his run-off roof observatory in the style of a Japanese tea-house (!). The engineer insisted on building it out of steel, and the whole structure became unnecessarily massive and expensive. A folding south wall was not possible, so Bill had to work around this by using windows. The structure between and above the windows obstructed a lot of sky, so then Bill had to work out a "ridiculously complicated" (in his own words) pier, allowing him to change the height of the telescope (which, however, also changes the polar alignment). Bill also says he ended up with insufficient space round the telescope, and a pointed roof that

---

[1] No. 17, October 2006.
[2] *More Small Astronomical Observatories*, Patrick Moore (ed), Springer 2002.

is unnecessarily obstructive when rolled back. The structure is, however, rather beautiful (and Bill jokes that it could double as an earthquake shelter).

In designing your observatory, look at the examples in this book and the many more available through the web links given here, and maybe consult local astronomers through a society. Astronomical newsgroups and forums on the internet can also be tremendously useful, providing a gateway to a vast amount of knowledge and experience that is freely shared in the amateur astronomical community.

CHAPTER FIVE

Techniques of Construction

It is possible for one person to build a perfectly good observatory from start to finish, with no help whatever, if necessary. Some of the skills required are: measuring and calculating lengths, areas and volumes, mixing and laying concrete and/or mortar, using an electric drill and electric saws, and understanding how to use screws, nails and bolts. More advanced carpentry techniques are useful, but not essential, and overall, this is not a particularly demanding list of accomplishments. The essential techniques, for those who do not already possess them, may all be derived from DIY books. Elementary safety precautions should be observed. Wear eye and ear protection when using electric saws, and gloves when using cement, mortar and concrete. Always make sure you wear nothing which could get caught in power tools. Do not deal with mains electricity unless you are well-versed in what you are doing. Leave this to qualified people.

Tools

Quality tools are always worth the money. Cheap tools will generally prove frustrating, inefficient, and possibly injurious. For working in wood and other soft materials, most of the tools needed will be simple hand tools: measuring and marking devices, spirit levels, squares, hammers and screwdrivers, hand and tenon saws, planes, rasps and chisels, pretty much covers it. While the project could theoretically be completed without any power tools, there is no particular sense in this today, as such tools are so inexpensive, and save so much labour. An electric drill is the most essential power tool, followed by an electric screwdriver, a circular saw or table saw, and possibly a jigsaw and electric sander. Obviously, the more complete a workshop you have, the better the results you will be able to achieve, and the quicker you will be able to achieve them, but these are the essentials. A reasonable workbench will be required, with jaws that can clamp pieces of wood. The larger and more stable this is, the better, though it may be folding. Very lightweight folding workbenches tend to be unhelpful.

Tools need to be sharp, and kept in good condition. The great enemy of hand-tools is rust, particularly when they are kept in garages or sheds. Rust can be kept at bay by giving cutting edges a thin coating of oil before putting them away. A piece of advice from the late Reg Spry[1] I think is worth re-stating. This is simply to always put each tool away in its correct place as soon as you have finished using it. If you can do this religiously, it saves so much frustration and time wasted searching for the appropriate tool.

Foundations and Stability

Once you have selected the site and drawn the plans, the next stage of the observatory project consists of levelling the site and creating foundations for both the observatory and telescope. These must be kept separate, so that there is minimum possibility of vibrations from the floor being transmitted to the telescope. Generally speaking, massive foundations are not required for an amateur observatory. If the lower parts of the structure are brick or block construction, then concrete foundations will be required, but these need be only under the walls, and probably no more than 15 cm (6 in.) deep. A concrete pad under the whole structure really is not required, and if a concrete pad is used for the floor it will be cold to stand on, it will offer no cushioning for dropped equipment, and it will retain daytime heat into the evening, damaging seeing conditions. Far better to lay a wooden floor over two courses of brickwork, or one course of blockwork, and have the telescope pier rising through a cut-out in this.

The lowest timber elements of the structure must be raised at least 15 cm (6 in.) above the ground. If the walls are timber, then concrete foundations are not really required, and something like pre-cast concrete stanchions, or fence-posts laid on their sides, and placed on a levelled base of sand about 7.5 cm (3 in.) thick, would provide an adequate base for the whole structure. With a timber structure, a certain amount of movement of the elements with respect to one another, with time, is normal, and can be accommodated.

It is possible to base an observatory on a pre-existing concrete or slab surface or patio, but it will be necessary to excavate this for the telescope pier. If a wooden structure stands directly on a flat surface that can absorb rainwater

---

[1]Reg was a most ingenious fashioner of telescopes and observatories from old bits of wood, as recounted in his book *Make your Own Telescope from Everyday Materials* (Sidgwick and Jackson, 1978).

across an adjacent open area, then that structure is likely to start to absorb water, and to rot, as it stands on a saturated surface. To prevent this it needs a damp-proof course under the timbers. Raising it on bricks would probably work as well. A telescope tripod or pillar stand placed on a concrete block on a pre-existing surface would be possible for non-critical, casual, observation, but for serious use of the observatory, and to get the best out of any telescope, a proper telescope pier, dug deep into the ground, is a must.

The digging-out and laying of concrete foundations is very hard labour, and, if no free help is available, it might well be worth employing labour for this part of the operation. Items can also be hired which save a lot of work in this department – specifically, a hole-boring machine and a concrete mixer. The quantity of concrete required for observatory and pier foundations is likely to be only a few cubic metres (or yards), or less than one, if only the telescope has foundations. This is a quantity that might not be economical to have delivered ready-mixed, but it is a quantity that is very hard, backbreaking, work to mix by hand, with a shovel. Perhaps the best solution will be to buy the ingredients, which are cement, sand and gravel (or the mixture known as ballast), or a dry concrete mix, in bags, and to mix and pour them using a small concrete mixer on the site. If the concrete is delivered ready-mixed (in other words, wet), it will be essential to have several people to help with moving it to the site and laying it before it dries.

It is hard to state the correct minimum size and depth of the foundations required for a telescope, as it depends on local geology, the environment, in terms of sources of vibration, e.g. roads and railways, and size of telescope. On a site with hard bedrock, like granite or limestone, and no particular sources of vibration nearby, excavating merely to the top of the bedrock, and casting a 60 cm (2 ft) square concrete block, would probably be enough for a moderate-size telescope (say up to 30 cm or 12 in.). On a site with soft bedrock, such as sand or clay, close to a major road, I would advocate excavating 2 m (6 ft) deep over an area of 1 m^2 (3 ft), which will probably need a machine, and casting a reinforced concrete plinth slightly smaller than the hole, then infilling around with sand, to further isolate and dampen vibration. This prescription may still not be adequate to ensure stability in some cases. It may be a good idea to consult a structural engineer with experience of the locality. Clay subsurface is a particular problem, as it shrinks and expands with changing groundwater levels, and also moves when water freezes. The aim is to have an absolutely stable platform for the telescope, which will have a completely unvarying polar alignment.

Some people have founded telescope piers in sand rather than concrete. This makes them easier to dig out and move at a later stage, and for a small instrument, say a short refractor of up to 150 mm (6 in.) aperture, might be adequate. If casting a concrete plinth, it should be decided at this stage what the telescope pier is actually going to be. It can be a cylindrical or square reinforced concrete extension of the plinth. This will be totally immovable in the future. It will be cast using some former, shortly after the plinth itself. Alternatively, it can be a steel pier, in which case, threaded rods or inverted bolts should be incorporated into the plinth when it is cast, to facilitate attachment. A steel pier, optimally, should have all vibration dampened by filling it with concrete; however, that again makes it immovable, so sand can be used instead. For a telescope below 30 cm (12 in.) aperture, a 15 cm (6 in.) steel column will probably be adequate.

Steel columns are generally expensive, and the cast concrete solution is much cheaper, but will need to be bulkier. I used, for the present main telescope pier in my observatory, two stainless steel bins placed on top of one another, as the moulds for the pier foundation and pier (Fig. 5.1). The larger is 37.5 cm (15 in.) in diameter, the smaller 30 cm (12 in.). The larger one was embedded in a larger concrete block forming the foundation, and linked to it, and to the smaller bin also, by steel pipes driven into the clay subsoil and running right up the column. An aluminium block was cast into the top of the column, to which the mounting base-plate is bolted. Alternatively in this situation, threaded rods

Figure 5.1. Telescope pier in my observatory, cast in concrete in two stainless steel cylinders (bins), with plywood shelf, and aluminium platform set in the top.

or inverted bolts (or J-bolts) could again be cast in, to attach the mounting. (As a post-script to the use of the stainless steel bins for casting the pier, they have not proved all that stainless in contact with the alkaline cement, and have started to rust.)

I previously tried making the telescope pier out of brick (Fig. 5.2). This pier worked, but it was bulky, and not so strong as the much narrower concrete and steel pier which replaced it. Laying bricks is time consuming and needs to be done quite precisely, and brick structures are in fact surprisingly elastic, but with little impact shear strength. Overall, I do not recommend a brick pier, though it looks quite nice.

A note on the subject of the levelness of pier tops: it is often believed that the top of a telescope pier needs to be perfectly level for obtaining a correct alignment of an equatorial mounting placed on it, but this is not the case. It should be as level as you can reasonably make it, but it is not critical. The idea seems to have arisen because portable mountings are often set up by a procedure which involves first levelling the mount with a bubble-level, and then setting its orientation. This is reasonable if you are relying on the angular setting of the polar assembly being correct with respect to the horizontal, but a precise setting of an observatory mount is not made in this way. It is made using the drift method (covered in Chapter 10). The drift method, performed iteratively with increasingly fine adjustments in azimuth and altitude, will give the correct result even with a non-level mount base. All that is needed, in the end, is for the polar axis to point at the pole. The way the rest of the mount lies does not really matter. A concrete

Figure 5.2. Earlier brick pier in my observatory.

pier top will not be perfectly flat, so, if bolting a mount base directly to this, it will probably be necessary to make a three-point support arrangement with washers or other small pieces of metal, to ensure rigid fixing without stress or bending.

It must be decided in advance how tall the pier should be. In general, Newtonian configuration telescopes have short piers, rising no more than 1 m (3 ft) above the floor level, and refractors, Cassegrains, SCTs, and similar, have taller piers, 1.2–1.8 m (4–6 ft) high. A Cassegrain-configuration telescope, particularly a large one, is awkward on a pier that is too short as, even if the observer is seated and using a diagonal, the observing position can be too low. The correct height for such a telescope located in a temperate latitude will be one that puts the eyepiece at a comfortable height for seated observation when the telescope is on the meridian and pointed at an altitude of 50–60°. German equatorial mounts create a large range of eyepiece heights depending on whether the object being observed is close to the meridian or not. Experimentation might profitably be made with an adjustable tripod, if the telescope and mounting are available before the observatory is built, to establish the optimum pier height. It is desirable to be able to avoid having to stand on ladders or platforms to reach any observing position that is likely to be used (but the telescope will not normally need to look right down to the eastern or western horizons). Long-focus Newtonians (above f6), greater than 25 cm (10 in.) aperture, probably will need the observing position accessed using steps or ladders sometimes, as will even short-focus Newtonians above 35 cm (14 in.).

In all cases, it should be ensured that neither the pier nor the floor obstructs the movement of the telescope to any part of the sky. It is also desirable, when dealing with a German equatorial, to arrange the pier to minimally obstruct the horizontal "throw" of the telescope, that is, to allow the telescope to revolve round the polar axis to be lower than the counterweight, for an hour's travel or more past the meridian (if the mount itself allows for it – some high-end computerised mounts do not). The advantages of this in both visual observation and imaging are considerable. It enables objects to be tracked across the meridian without having to reverse the telescope and counterweight. Since the meridian is the ideal place in the sky at which to be observing any object, it allows images to be taken at optimum altitude for an extended period, without this annoying interruption. Even for visual observing, the benefits are worthwhile, of not having to turn the telescope over so much – an operation which upsets collimation and finder alignments in most telescopes with mirrors.[2]

Materials and Joining Them

There are many possible methods of observatory construction, and many possible ways of using the materials. It will not be possible to cover all possibilities

---

[2]Many 19th century German equatorial mountings were made with a massive cast-iron sloping pier which lay approximately parallel to the polar axis, and allowed the telescope to go right round the sky without colliding with the pier, totally eliminating the reversal problem. This pattern is never manufactured today.

here. Nevertheless, I will offer some guidance here on the use of the most likely structural materials, which I hope will be of use to those who have not tackled this kind of project before. DIY experts might skip this section.

You may decide to make parts of your observatory permanent, using brick or block construction. You can either employ a bricklayer, or learn the technique of mortaring and laying bricks and blocks yourself. It is not terribly difficult, but requires a fair amount of practice to get it right. Even if the results do not look terribly neat or completely professional, they may suffice. The brick pier shown in Fig. 5.2 was built with no previous experience of bricklaying. I also built supports for the timber rails for my run-off roof out of brick. This keeps them completely rigid. The basic rule for bricklaying is to keep everything level, levelling each course of mortar and each brick with a spirit level in both directions before going on to the next.

For most people, wood is the easiest material with which to work, being easily-available, fairly cheap, and easy to cut and drill. The term "wood" embraces: external grade timber, which is not dried, not planed flat, and treated with a preservative which gives it a slightly greenish colour; internal grade timber, which is dried, and may be unplaned or planed, but is commonly planed in the smaller sizes, and is not treated; plywood, which is strong and versatile, but needs to be kept dry; chipboard and fibreboard, which are most useful for flooring and roofing purposes; and hardboard, which is weak, and mostly useful for lining and insulation.

When cutting timber and sheet materials, remember the basic maxim: measure twice before cutting once. A cut cannot be undone. For accurate cuts on small timbers, use a tenon saw. A general flexible saw is used for larger timbers and sheets. A circular saw saves a lot of effort when cutting substantial timbers, such as floorboards, where accuracy is not critical. Shapes can be cut into boards using either a padsaw or a coping saw (two types with narrow blades), or an electric jigsaw.

Timber sizing is most confusing. Pieces of timber are sold by timber merchants with a "nominal size" for the cross-section that is several millimetres (or up to an eighth of an inch) greater than the actual size. For example, so called "2 × 2" in. timber is actually about 45 mm^2, rather than the theoretical 50.8 mm^2. This, supposedly, had to do with the way timber used to be planed, in a way that its size turned out slightly unpredictable, but I am not sure why it persists, and in fact many stores that sell packaged timber now quote exact dimensions.

The "2 × 2" would be a reasonable weight of timber for most of the framing of a typical observatory, though for a large observatory one might go to larger sizes. For my small run-off shed (approximately a 1.5 m (4 ft 11 in.) cube) I used mostly 2 × 1 to keep the weight down, and this has been satisfactory. External grade timber, though rough in finish, is usable in the framing of observatories, and for their cladding, and, used inside, it is less likely to rot through leaks and condensation. Its lack of sizing accuracy makes it difficult to ensure accuracy in construction, however. Planed wood is better for accurate jointing.

If new to working with wood, you will need to understand its basic properties. Fixings into wood cannot hold in the end grain. Therefore, there are two sides of a piece of wood that can be used for fixing, and one which cannot (Fig. 5.3, A). This leads on to the whole subject of jointing. It is, however, not necessary to get

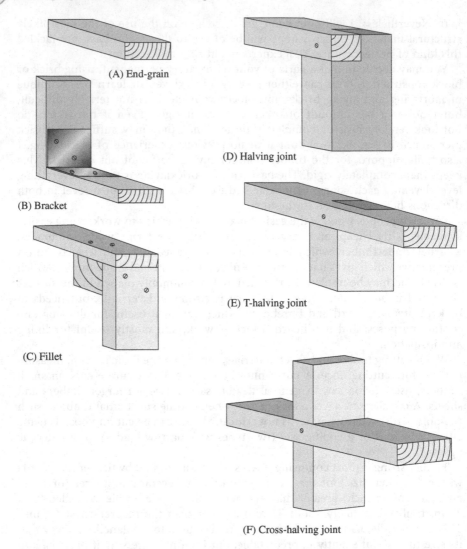

Figure 5.3. Simple methods of joining timber with screws, avoiding fixing into the end-grain.

into sophisticated joinery in the building of an observatory. There are two simple methods by which timbers can be joined at right-angles, or other angles, without joints. These are the use of metal brackets and wood fillets (Fig. 5.3, B and C). If you can use a tenon saw, then you can make halving-joints, which are better (D). If you can use a chisel as well, you can make T-halving and cross-halving joints (E and F). A strong framework can be constructed using these three joints, reinforced with woodscrews.

I strongly recommend all important structural members to be held together using screws rather than nails. Screws are far easier to remove, if necessary, than nails, and hold better. Removing nails usually causes damage, or proves impossible. Screws will come out easily, especially if lubricated before use with grease, oil or soap. Screws exert a strong clamping force to bring parts into line. Woodscrews have a threaded tip and a smooth shank, and can be countersunk, or round-headed. The countersunk type are the most useful. The holes for them are drilled in three stages: a pilot hole through both pieces of wood, into which the screw thread digs; a wider clearance hole through one piece of timber, through which the thread passes, and in which the shank rotates; and a countersink hole to make the head flush with the surface, or sit slightly below it. Screws that are used to affix pieces of metal or plastic to wood are best of a totally threaded design, to obtain maximum grip.

The heads of screws are either slotted, or have a cross in them (Phillips or Pozidriv heads), or occasionally a square hole, or other shaped hole for a special bit. The slotted type is traditional, but I recommend against these. They have no advantages in use. The other types are all better, and work better with an electric screwdriver, which I also recommend as a great labour-saver. I also recommend against brass screws and brass fittings in general, as they are weak. Try to buy high-quality stainless steel screws, bolts, nuts, and fittings. If you pay for high-quality fittings and fixings, you will find it is worth it in terms of ease of use and the saving of frustration. There are even more corrosion-resistant "external grade" fixings available, which sometimes have a green coating. These are worthwhile for places that will be exposed to continual dampness, such as those outside near the ground level.

Though screws are fairly easy to remove, they are not so easy to remove, and not so strong, as nuts and bolts, and, though expensive, I recommend these for use if it is anticipated that the structure might need dismantling in the future. If this is the case, you should plan your observatory so it can be disassembled into a small number of sections easily (and not later tie these together with wiring or other impedimenta). These sections can be held together with a small number of large coach-bolts. Coach bolts have a square section under the head that prevents them from turning in a hole drilled in wood, and they are used in heavy timber construction. They should be used in combination with heavy washers, to allow the structural elements to be clamped strongly without damage. Nails should only be used for relatively superficial elements of the structure, such as trimmings and claddings. Nails are a timesaver, if strength of fixing and demountability are not needed.

All fixings are very expensive if bought from DIY chains. I recommend ordering them in bulk from mail-order suppliers, if you do not want them to exceed the total cost of all the other observatory materials.

An alternative to timber-framing is the construction of a shed using a welded steel framework. This is not so daunting as it might sound to the novice. Magnesium inert gas or MIG welding is by far the easiest type of welding to learn, and an MIG welder can be bought for less than £200 ($350). MIG welding uses a small cylinder of carbon dioxide to deliver a gas shield that prevents oxidation of the components while welding. You must read-up on safety before embarking: a welder's headshield or helmet, leather gauntlets and a leather apron must be worn. Protection of the eyes is critical.

Square-section, L-section and channel-section steel is obtained from a steel stockyard. It may be cut with an angle grinder, but this is slow, and it is best to order it cut to the required lengths. The angle-grinder is, however, still required, to grind the steel clean prior to welding. If you have not done any welding before, a supply of scrap steel will be necessary, on which to practice; you will get through quite a few wire-feed tips while learning.

Simple overlapping joints are the easiest to make, and also the strongest (Fig. 5.4). A large builder's square is needed, for getting the frame square, plus G-clamps or welding clamps (similar to mole-grips), for holding the members in place during welding. Note that, when using components such as the steel castor shown in Fig. 5.4, all galvanising has to be ground off before welding. If welding you own frame does seem too daunting a task, the job could easily be taken on by a garage or agricultural mechanic, if given a design. This could be quite a quick and cheap method of obtaining a strong observatory frame.

There are numerous possibilities for cladding the framework of an observatory, if the sides are flat. One is plywood, which is very strong, perhaps unnecessarily strong, heavy and quite expensive. In addition, it will de-laminate easily if it gets wet, so requires very good sealing using varnish or paint, and very good maintenance of those coatings. The best type is "external-grade" plywood or "marine ply". My first observatory was clad with hardboard, but I do not recommend this, as hardboard will warp when used outside, even if painted. The best and most attractive claddings, in my opinion, are featheredge timber and shiplap timber. Garden sheds are normally clad with one of these, shiplap being reserved for the more expensive models. Hence, an observatory so clad may

Figure 5.4. Jointing in a welded-steel run-off shed framework.

easily be mistaken for any other shed. The cladding will take on the colour of the wood preservative that is applied to it over the years. This is normally a brown creosote substitute (creosote itself is now banned), but coloured wood-preserving stains can also be used.

Featheredge timber is the material usually used for fences. It is fairly rough, and has one edge narrower than the other. Assembling a featheredge wall or panel is fairly time-consuming, but one acquires the technique quickly. The boards need to be laid so that the thick edges point towards the ground in use. They are overlapped, so the thick edge at the bottom of one board overlies the thin edge at the top of the next one down, with an overlap of about 2 cm (3/4 in.). It is as well to measure this overlap and mark it on all the boards with pencil before commencing nailing them to a frame. Several boards should be laid in place to get the pattern, before the first one is nailed, as nailing the first will also fix the second, and so on. Featheredge construction is cheap and attractive, allows the shed to breathe, and keeps water out efficiently. I have used it very successfully for my two present observatory buildings.

Shiplap timbers are more carefully made than featheredge, profiled to slot into one another. They thus make a waterproof construction relatively quick and simple. The timbers are just cut to length, slotted together, and nailed to the frame. They tend to shrink over time as they dry out (as do most timbers, even when used outside), and this can reduce the keyed overlap in time to the point where gaps occur. This is minimised by good maintenance with preservatives. Shiplap constructions look professional, and weather well. Both featheredge and shiplap claddings are normally finished at the corners of the shed with a moulding or beading, a narrow strip of wood nailed into the corner, for a neat effect.

An alternative to assembling panels of overlapping timbers is to buy the panels ready-made, as fencing panels. This, potentially, can save a lot of nailing, but the quality of construction will not be brilliant; they are usually rather flimsy, and the size of the panels available will have to determine the dimensions of your observatory. There have also been observatories built from old doors. When you come to look around, it is surprising how often you find that valuable, usable doors have been discarded. However, you would need quite a lot of them. Constructors of amateur observatories tend to be experts at finding and recycling this kid of detritus. A steel framework would probably be clad by bolting to it pre-assembled timber panels or plywood sheets. With the latter, this method of construction is very quick.

Roofs and floors can be made of individual planks of wood nailed to battens (for roofs) or joists (for floors), or they can be made of sheets of chipboard or fibreboard, which are cheaper per unit area, and quicker to put together, though heavy and not very strong. Chipboard sheets are sometimes made with tongue-and-grooved edges, so they are slotted together for extra strength in construction. This would not be a good idea in laying an observatory floor, as it is wise to make the floor easily removable in sections, to allow work to the foundations of the telescope, should that become necessary. Floor joists should be at least 7.5 cm (3 in.) high, to allow air circulation under the floor, to combat damp. The joists themselves should be well-treated to resist damp, and supported on brick or concrete above the soil. They should be spaced not more than 50 cm (20 in.) apart.

In creating a waterproof roof, the main choices are plastic sheets of various sorts, either corrugated or flat, affixed to a timber frame, or bituminous felts which can be used to seal solid timber, plywood or chipboard roofs. Plastic sheets are lightweight and thus suitable for moving-roof observatories. They are usually drilled and screwed down with special screws with waterproof plastic covers, but some systems involve nailing with special twist nails. Small holes should be drilled in the plastic for the nails, or else nailing may split it. Either way, if the sheet is corrugated, the fixings must always be put in the ridges of the corrugation, not the valleys. Obviously, the valleys must run down the slope of the roof to direct rainwater off. Flat multiwall polycarbonate sheeting, often used for conservatories, is stronger than corrugated plastic, but much more expensive. A disadvantage of all these plastic coverings is that they are usually transparent or translucent, so can create a lot of heating of the observatory in sunny weather. They can be painted with white or aluminium paint to combat this, or lined inside with reflective plastic sheeting.

Roofing felt is cheap, and can be laid over any kind of board. It is secured by folding over at the edges, and nailing into the side, or underside, of the board, with special large-headed galvanised roofing nails. Joins are waterproofed using black mastic, and the sheets are arranged with the "uphill" sheet overlying the "downhill" one, so water cannot flow into the joins. It is not necessary to nail the joins. The method of using a torch to melt bitumen to fix down roofing felt, as used on flat house roofs, would be excessive for an observatory, and needs to be carried out by a professional, anyway.

At least one observatory has been roofed with tiles, but that is bizarre. Aluminium sheeting is good, but very expensive. Any roof needs to overhang the walls significantly at the eaves, with the largest overhang, for a pent roof shed, being on the "downhill" side, so that water that is shed off the roof falls a few inches away from the foundations, and does not splash back onto the underside of the floor. In certain weather conditions this will probably still occur, but it should be minimised with an overhang of at least 15 cm (6 in.). Guttering is difficult to arrange for a moving-roof observatory, though it has been done; a run-off shed is usually too small to need it.

One other point worth making about flat (pent) observatory roofs in run-off roof designs, because it is not an obvious one, is that if they have a jointed timber framework, it should be borne in mind which way the force of the weight of the roof is going to act on the joints, to avoid a sagging roof. The roof will have little tendency to sag in the direction in which it rolls, as the sides running in this direction will be supported by wheels. In the other direction, however, the cross-beams will be acting as lintels, resisting a bending force. The members need to be jointed so as to put the joints into *compression*, in order to achieve a rigid roof. If it is done the wrong way (as I did with my observatory), the weight of the roof will cause the joints to *spread*, and the roof will bow. Fig. 5.5 illustrates this point. I had to reinforce my roof with extra timbers bolted to the ends to cure the bowing. With a metal framework this would not be an issue.

Doors are best constructed by the traditional method of making outbuilding doors, which is using keyed tongue-and groove timber, ledged at top and bottom and diagonally braced with plain timber inside (Fig. 5.6). Hanging doors on hinges is a bit of an art. They need to be supported in their final position when the

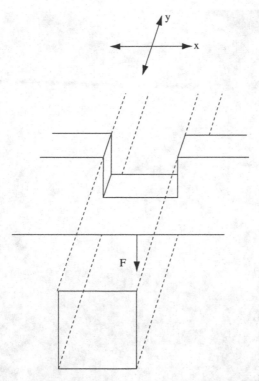

Figure 5.5. Illustrating the point about jointing in a run-off roof framework. If the roof runs in the x direction, then the dashed member will act as a lintel, and the downward force F will cause the joint to spread and the roof to bow. If the roof runs in the y direction, the solid member will act as a lintel, the joint will compress, and the roof will not bow.

hinges are screwed, but will always fall back slightly, a fact that needs to be taken into account. Also, it should be realised, particularly when dealing with anything like a door or a moving roof, which needs to open and close fairly precisely, or when attaching fittings such as bolts, catches and locks to timber elements, that anything made of timber will expand or contract, shift and adjust, with time, and that the relationship between the parts will not remain fixed to better than 8 mm (1/3 in.) for very long. This continual adjustment of wooden structures is in their nature, so elements that open and close cannot be made to fit too precisely, and fittings will often have to be moved later. A steel or aluminium-framed building would be more stable and precise.

No mention has been made so far of fibreglass construction. This is, without doubt, a messy business, if you mould the fibreglass sections yourself. However, it is the only way for the amateur to make lightweight, curved sections for a cylindrical building, and probably the only way for him to make a truly hemispherical dome. The basic technique is to create a mould, which can be wood, papier-mâché, cardboard, or wire mesh over a plywood frame, and then to lay over that alternate layers of glass fibre matting and resin, to the required

Figure 5.6. Tongue-and-groove construction, ledged and braced door on a run-off shed.

thickness. Each layer needs to be thin, otherwise cracking or warping may occur on hardening.

When the fibreglass has hardened, the mould is removed (or it can remain in place, in the case of wire mesh). The chemicals involved in the resin are noxious, and the operation needs to be done in a very well-ventilated place. The temperature should be quite low, or the resin will set too quickly. Its rate of setting can be varied to a certain extent by adjusting the quantity of catalyst. The materials for making fibreglass are not cheap, probably coming to at least £500 or US$800 for a decent-sized dome (without the walls), and, bearing in mind all the difficulties in doing it yourself, it may be better to have the layering done on your mould by a professional with dedicated facilities. People who make truck bodies or boats will be able to help. If the construction is designed in fairly small sections which bolt together, which is a good idea anyway, then the moulds and finished sections can be transported around. If a dome were to be made in one piece, however, the work will probably all have to be done on site.

It is not completely necessary for a fibreglass observatory to have a continuous timber or metal frame, though some do. The sections can be made strong enough to be self-supporting when bolted together, through techniques such as embedding timber or metal bracing in the fibreglass layering. A disadvantage to fibreglass is that, in its natural state, it allows in quite a lot of solar radiation, causing problems with daytime heating. Again, this can be countered with paint. Aluminium paint might also help with the other problem of fibreglass constructions, their highly electrically-insulating nature, which can cause problems of static build-up, and even lightning strikes, in exposed locations. More will be said about these points in some of the case-studies in Chapter 9.

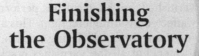

CHAPTER SIX

Finishing
the Observatory

Paint and Preservation

Timber structures can last a lifetime, but, if they are to do so, they need close attention to prevent dry rot, wet rot, mould or fungus from setting in. The preservative that is already applied to external treated timber provides a basic level of protection, but it is not sufficient for timbers that frequently get wet. Probably the best rule for timber is, if in doubt, to apply more preservative. There are many kinds, varying widely in price, and also chemicals that attempt to restore decayed timber, though it is best to avoid getting to that stage, and additionally, to build in such a way that parts which are more likely to decay (such as the flat timber rails in my run-off roof design) are easily replaceable, should that become necessary.

Essentially, you get the level of protection that you pay for, but for timbers that are not continuously in contact with water, a basic preservative applied every 2 years should be sufficient. The standard wood preservative of the past was creosote, but that is carcinogenic, and became illegal in the European Union in 2003. Its substitutes go by names such as Creocote, or just "creosote substitute". These chemical mixtures generally turn timbers a mid-brown, which darkens with further applications. It would be more desirable to keep an observatory building light in colour, and hence cool, so transparent treatments would be better from this point of view. Creosote and its substitutes are applied by brush or dipping. They kill plant and aquatic life, so should be handled with care. Brushes are cleaned in white spirit.

More benign are the water-based outdoor timber treatments, which contain an emulsion of wax, are more expensive, and come in a variety of colours. These work best on rough, sawn timbers. Planed timber is better painted, but it can also be creosoted, though it does not absorb the treatment so well as rough timber. Pretty much all the timbers in an observatory should be treated, even if they are inside – they will still most probably come into contact with a lot of dampness through dewing.

Timbers can also be varnished, but all varnishes perform poorly out of doors, where they are adversely affected by ultraviolet light. This applies even to external or "yacht" varnish. Polyurethane, or quick-drying varnish, cannot be used externally, as it dissolves in water, but all the varnishes crack up after a couple of years and need to be sanded-off and reapplied, so I don't recommend them for external observatory preservation purposes. They can be well and attractively used internally. Timber decking need special attention, as water lies in the channels moulded into the boards. There is a wide variety of special sealants and oils manufactured for the treatment of decking.

Metal surfaces also need painting, generally after priming, and repainting as necessary. The best primer for ferrous metals is the red oxide type. Plastic and fibreglass need not be treated in any particular way, merely monitored for integrity. They can be painted with ordinary finishes, and priming or undercoating is not necessary.

Some people paint or otherwise coat the inside of their observatory. This, obviously, helps with preservation of timber, and, particularly with run-off roof observatories in towns, can be used to darken the interior so it reflects the skyglow less, to aid in dark adaptation.

Flooring

I am a keen advocate of wooden observatory floors, as they don't absorb or store much heat, are pleasant to use, and work well, provided they are well-ventilated underneath, and isolated from the telescope pier. However, a concrete floor would be much improved by adding a layer of carpet (you will find that perfectly serviceable old carpets are always being thrown out), or, better, carpet over some kind of boarding. This will insulate the floor, making it more comfortable to stand on, and lessening thermal problems. But the main benefit will be the protection given to valuable eyepieces, filters, and cameras, when they are dropped. Again, a dark floor covering will darken the inside of a run-off roof observatory substantially.

Regulation of Temperature and Humidity

A response to these issues obviously will depend very much on your local climate. The requirements are that the observatory and its equipment do not get too hot (I cannot see any harm coming from extreme low temperatures per se), that all elements of the observatory can reach outside temperature as quickly as possible when required, and that the structure and equipment speedily dries out after it gets damp.

Some amateurs in hot climates have taken a leaf out of the professional book and installed air-conditioning in their observatories, so that when they are opened at nightfall the inside temperature is already close to that outside. In the UK and most of North America this will not be necessary, but for domes, fans are a very good idea, particularly in the summer. The best seeing in north-west Europe generally occurs in spring and summer, when the North Atlantic jet-stream (a high-altitude wind) is least active. Lunar, planetary and double-star observers, particularly, using domes, will probably get the best results at these times if they ensure the air is well-circulated, avoiding the "chimney" effect, where warm air continues to leave the observatory through the slit long after it is opened. The worst summer thermal issues tend to occur in metal domes, which get hot, and concrete-floored observatories, which have very high heat capacity. The combination of the two tends to be fatal to high-resolution work.

Observatories that are adapted from commercial sheds or summer-houses will probably have too many windows, and behave like greenhouses. The only real remedy would be to paint the windows out, or make wooden shutters, though blinds might be partially effective. Attempts to make specifically-designed observatory buildings more "attractive" can be another reason for the occurrence of thermally-unhelpful glazing. Fibreglass domes are usually translucent, and can get rather warm, though they could be painted with white or aluminium paint, and the same goes for plastic-roofed shed observatories. Dark colours are worse than light ones, but dark colours, particularly green or brown, are often desired, or specified, by non-astronomical authorities, to make an observatory blend in with the countryside, unfortunately. In this case, the best the astronomer can do is to try to ventilate as well as possible.

The telescope design is also a factor in the extent to which local warm and cold air currents will affect seeing. The closer these currents are to the objective, the worse the results will be. Refractors and catadioptric scopes, being closed, will generally do better than reflectors in a thermally un-equilibrated observatory, where warm air will tend to rise from low down, and skeleton-tube or tubeless reflectors will be the worst of all. A 0.1°C difference in the temperature of the air in different parts of the light-path of a telescope will create a difference in effective path-length for the light greater than a quarter-wave – in other words, it will be greater than the minimum tolerance for mirror figuring, effectively ruining the telescope. In general, deep-sky observation is much less stability-dependent than lunar, planetary and double star work (though the effects of poor seeing will certainly be seen on globular clusters and planetary nebulae when they are examined at high power), and, for an observatory used mainly for deep-sky observation, thermal issues are likely to be less important. In all observatories they become less important in the winter, when diurnal temperature variation is usually small.

In temperate climates, if the problem in summer is overheating, in winter it is usually moisture. The most afflicts run-off roof observatories, where the whole interior of the structure, when exposed to the night, cools radiatively to below the dew-point, and becomes wet. In a full run-off shed observatory, at least only the telescope becomes wet. This dampness becomes shut-in when the observatory is closed, which does neither the observatory, nor any of the equipment inside it, any good. Fans might again be usefully employed in such cases, or a small electric convection heater left on at a low level, or a mains incandescent light-bulb. Some amateurs have installed dehumidifiers in their observatories, or wall-mounted extractor fans. Some very cheap, small, mains-powered dehumidifiers, made for use in small spaces such as cupboards, are now available, and I have found these very effective in my observatory sheds. The buildings and the optics are no longer wet the day after a winter observing session, as they were before.

Security

Security will be a matter for personal judgement, based on the vulnerability of the location. As I have stressed before, an observatory that looks like a garden building is unlikely to be suspected as a home for valuable equipment. On the other hand, thieves will sometimes try to get into outbuildings with a view to obtaining tools as an aid to burgling a house. Urban observatories are likely to be less vulnerable than rural ones, as they are generally located in gardens, or on plots which are well hemmed-in by fences, hedges, trees and other properties, and not very visible from outside the property. An observatory in the middle of a field or on top of a hill is much more obvious.

Societies owning observatories have a lot of difficulty with security when their site is not on anyone's residential property, and particularly when it is isolated. In some cases a burglar alarm linked up to the police might be wise. Town dwellers might also wish to fit an alarm. Again, avoiding having windows is a good idea. Another point is to be careful, in your possible enthusiasm for advertising your astronomical achievements and acquisitions on the internet, not to provide information which will allow your exact location to be pinned-down by a criminal (and worse, allow him to work out on which date you will be out at a star-camp or party). It is really a matter of common sense.

Unfortunately, telescopes that are left outside all the time are quite often stolen. They are sitting targets. In terms of other equipment, the amateur astronomer is in a relatively good position, in that the very high value of much specialised equipment is not likely to be known to the average criminal, and, in any case, he would have difficulty in realising that value, because the astronomical world is quite small, there are relatively few people who might buy the equipment, and word can quite easily be put out on internet forums and newsgroups to warn other astronomers to be on the look-out for equipment which has been stolen. Obviously, prevention is better than cure, so the basic rule is to lock everything as thoroughly as seems sensible when no-one is in attendance. Even an expensive lock will only cost a tiny fraction of the cost of the camera or top-quality eyepiece that it could help you to hang on to.

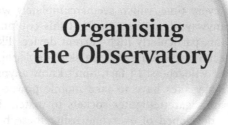

CHAPTER SEVEN

Organising
the Observatory

A well-organised observatory will make observing more of a pleasure, and more productive. Much how this is done will be down to personal choice, the type of observing that is to be done, and the choice of equipment. Here I will deal with some of the main internal issues in an observatory.

Electricity

It is sadly not possible to get very far in astronomy these days without an electrical supply. Gone are the days of falling-weight and clockwork telescope drives. Everything seems to need its own electricity supply these days, and many of these things seem to need different voltages.

Most telescope mounts require 12 V DC, but some, particularly Meade equipment, take 18 V, and some, even, 24 V. Old mountings using synchronous motors require 240 V AC if they are British, or 110 V AC if they are North American, but these are best (most safely) supplied using an inverter, which generates this voltage from a DC source, normally 12 V. In this case, they are still less safe than low-voltage equipment, and the whole telescope and mounting must be properly earthed. Overall, they are not to be recommended, and conversion to DC motors is a good idea, if possible.

Much equipment that uses 12 V DC is provided with plugs of the car cigarette-lighter type, which is most unfortunate, as these connectors are really unsuited to critical applications. They are unreliable, as they have poor contact with the socket, particularly at the tip, and become disconnected far too easily. This can be a major frustration if your driven mount becomes disconnected half-way through a one hour exposure of a faint galaxy, or your GOTO mount loses power and has to be re-aligned.

This situation seems to have come about because manufacturers have imagined that everyone will be going into the backwoods in their cars with all their equipment to observe every time, which seems completely wrong, in principle and in practice. And anyway, if you use much of this equipment connected to your car battery for long, particularly high-current devices like mounts, laptop computers, and dew heaters, you run a severe risk of depleting your car battery, and getting stuck in the wilderness! In fact, don't know anyone who does this. Those who go to dark-sky sites have to take mobile power-packs with them, which are equipped with cigarette-lighter sockets to match the equipment. So a barmy situation has developed of everyone using these bad connectors. In advance of the manufacturers seeing sense and coming up with something better, I recommend all these 12 V plugs and sockets be cut off, and replaced with something better. Screw terminal blocks will do, though these are not so easily disconnected.

In an observatory, it is possible to use batteries, but from all my I experience, I strongly recommend against it. It is a chore to have to keep recharging batteries. If they are high-capacity, they will be heavy, and will have to be carried to a mains point, or a cable will have to be laid from somewhere else temporarily to facilitate charging. Small batteries will always be running out at the worst moment, when conditions are perfect, and you are just about to bag your prize image. All batteries are prone to lose voltage badly at low temperatures, which are just the conditions in which you are likely to need them most. Large batteries are also bulky. You can do without the unnecessary problems created by batteries in an observatory.

So what is the alternative? It is often not that easy to lay a mains supply to an observatory some way from habited buildings, so there can be a certain reluctance to bite this particular bullet. I tried, for a time, a system of having a power supply in another outbuilding (a garage closer to the house, that already had a mains supply), and using a 12 V feeder wire from that to all the devices in the observatory. The trouble, I found, with this, is that with low DC voltage, power losses in a long cable are large, and with several power-consuming devices at the end, the power supply would have to be unusually large, otherwise there would be a risk of loss of adequate voltage to something critical, say, the telescope drive, when something else that draws a high current, say a dew heater, is connected.

I concluded it is far better to take the mains voltage to the observatory, even if that involves considerable work and expense, and to have there several independent power supplies producing the various low voltages required by different devices, so that there is no chance of the switching on of one device affecting anything else. Reliability must be the prime goal of powering observatory equipment. Of course, if there is a mains power cut, as happened one night during a star-party at my observatory, then you are "up the creek" again. But that has not happened with sufficient frequency to lead me to make contingencies for it.

North American 110 V AC is much less dangerous than European 230 V AC, but more susceptible to losses if run in ordinary cables over any distance. However, it must be stressed here the extreme danger of tampering with any mains supply if you are not fully conversant with what you are doing. There are, moreover, complicating legal aspects to installing outdoor mains electricity, and these vary between countries, and in the USA, from state to state.

In the UK, it is now illegal for new mains electrical circuits, as will probably be required for a new observatory, to be installed without certification. This means either hiring a qualified electrician to do the work, who will then certify that it has been performed according to Institute of Electrical Engineers (IEE) regulations,[1] or, doing it yourself, and paying for someone from the local authority to come and inspect it to see if it complies with IEE regulations. Either way, this increases costs. The official reasoning is to protect subsequent owners of the property from being put at risk by badly-performed electrical work, though this is hardly likely to apply in the case of observatory installations. It means work must be carried out by someone with a detailed knowledge of the regulations, which cannot be duplicated here.

In the United States, electrical work on an outbuilding such as an observatory must be in conformity with the state electrical code, and this conformity will be checked by the same code official who will check the code-compliance of other aspects of the building. The broad outlines of what is required are similar in the UK and USA and most other countries, but it is essential to check the legality of any mains electrical work you plan with your local authorities. However, it seems reasonable to state some general principles here.

There must be sufficient capacity in the main electrical service to the property to feed the observatory, and a separate circuit to the observatory is normally required, with a local panel to enable emergency cut-off in the observatory. Local laws vary on whether the supply needs to be underground or not. Trailing wires or unprotected wires slung at a low level should not be used. If they are trailing they are a trip hazard, and if slung, they can get accidentally cut. UK regulations stipulate that an overhead cable has to be at least 3.5 m (12 ft) above ground level over pathways, and at least 5.2 m (17 ft) above driveways. The maximum allowed unsupported span is 3 m (10 ft), but, if supported by catenary wire, it can be of any length. Alternatively, cable can be run overhead in continuous, rigid steel conduit, at a minimum height of 3 m (10 ft) above a pathway, or 5.2 m (17 ft) above a driveway. The conduit should be earthed.

Unless your observatory is very close to your house, running a cable overhead by these methods is unlikely to be easiest method. It is likely to be easier to bury the cable. This can pose a problem if the cable has to pass under concrete. If planning a new path or patio, it is very wise to incorporate a cable conduit in the construction. If a cable goes under the earth, it should be buried at least 50 cm (1.6 ft) deep, or deeper if it has to pass under areas that are likely to be dug. Ideally, a trench should be dug and given a floor of sand. Bricks should then be laid on either side of the cable, and paving slabs should be placed on top

---

[1] The body is now called the Institution of Engineering and Technology, but the regulations are still known as the IEE regulations, just to confuse.

of these to form a continuous protected housing. The slabs should be marked with black and yellow striped tape to serve as a warning to anyone encountering them. Finally the earth should be in-filled back on top.

Ordinary PVC cable can be used overhead, but is not permitted to be buried. For this, you need either armoured cable, which is protected by a layer of steel wire on the outside, or mineral-insulated copper-sheathed (MICS) cable, which contains magnesium oxide powder insulation within a cooper sheath, and must be used with special moisture-seals and junction boxes. Both these options are expensive, but the former is easier and more common. However, the cheapest option is to use ordinary PVC 3-core cable buried in plastic conduit. PVC insulated cable can also be run through plastic or metal conduit mounted on a wall. Flexible conduit can be bought which consists of a steel spiral encased in PVC, and this is a simpler, and strong, alternative to making a rigid conduit go round corners in the normal way, by cementing elbow joints onto straight sections. An electrician will advise on suitable methods of running outdoor cable for your particular situation.

UK regulations state that at the house end of the system, the cable must be connected to a switch-fuse unit and then to an RCD (residual current detector). The RCD is a circuit-breaker, which turns the supply off if a leak to earth is detected. This is the most important safety-feature in any outside wiring system. Plugs can be bought which have an inbuilt RCD, but UK regulations specify there must also be an RCD permanently wired into the supply to any outbuilding. The RCD is connected to the meter and to the house's main earth connection, but these connections must be made by the electricity company. The cable should pass from the outside, into the house, through plastic conduit, entering above the damp course, and, if possible, below the floorboards.

At the observatory end of the system, there should be another switch-fuse, and then, probably, a junction-box leading to lighting circuits and sockets. These requirements are summarised in Fig. 7.1.

Further details of electrical safety will depend somewhat on the design of the observatory. In a run-off roof observatory, or a run-off shed, the sockets must be weatherproof, at least IP55 rated. (The letters IP refer to a level of "ingress protection" for equipment, recognised in the European Union. The higher the number, the better-protected the equipment.) In a dome, arguably, the sockets

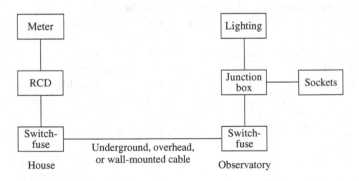

Figure 7.1. Schematic diagram of mains electricity supply to an observatory.

need not be waterproof, as there is much less condensation in a dome, but it would be as well to make them at least IP55 rated here as well.

In an observatory with fixed walls, the sockets and waterproof switches controlling them will be mounted on the walls, but this will not be possible with a run-off shed or temporarily-covered telescope arrangement. In this case, the wiring can be run under the floor, and sockets and switches attached to the telescope pier. They could be on the telescope pier in a fixed-wall observatory as well. If there is mains voltage anywhere near the telescope, as is necessary with synchronous motors, it is essential to earth the telescope and mounting. If the observatory is far from the house (say more than 20 m or 60 ft), it is probably as well to create a local earth with a copper and steel earth rod in the ground, and to connect the earth of the observatory supply to this, additional to it being connected to the earth of the house. Again, all these details should really be checked by a professional electrician.

Install enough sockets. You cannot, certainly in an open observatory, use normal socket extension strips, as these will not conform to outdoor electrical requirements, and will be dangerous, undoing the safety measures already built in to the system. I recommend calculating the number of sockets you think you will need, and then multiplying it not by two, but four, for realistic future-proofing.

Dampness getting into mains electrical equipment can prove fatal, so, if in doubt, do not use a mains appliance in an observatory, or get advice. Laptop computers in general are safe, because they have a well-sealed power supply feeding a low voltage to the computer. A desktop computer could be used in an open observatory if shielded by some sort of box loosely covering it, protecting it and the monitor from condensation and possible rain. It will probably generate enough warmth to keep itself dry. Desktop computers are definitely less safe in the open than laptops, however, and also less reliable. Mains hairdryers, which are sometimes used to clear dew from optics, are particularly unsafe, and should be avoided. The switches in them are poorly-protected and can shock in the damp. Better to rely on 12 V dew-strips or 12 V hot-air blowers, which can be bought as an in-car accessory. Remember, any mains switches or connections which are not well-shielded from damp ingress are dangerous in the observatory. Placing connections under a table or a shelf provides additional protection from damp, but they should still be at least IP55 protected.

Low-voltage power supplies that run off the mains, and power telescope motors, dew heaters, etc. are sold by the astronomical suppliers, but they are usually high-priced and not very powerful, and similar, and more powerful, units can be bought more cheaply from general electronics suppliers. Telescope manufacturers will tell you that you must use their power supplies to avoid invalidating your warranty, and one can see why they do this – they must get people doing all kinds of silly things. But generally all you need is a regulated supply that provides enough current (measured in amps or A) for all the things you need to run off it.

A regulated supply is different to the unregulated supplies also available, which should not be used to power expensive equipment like telescope mounts, which could be damaged by unregulated voltage. The 13.8 V high-capacity regulated supplies sold by electronics suppliers, made mainly for operation of amateur radio equipment, are ideal. They may have cigarette-lighter style sockets, or post terminals, which are better, or both. The figure of 13.8 V derives from the fact that this is the maximum voltage a (nominally 12 V) car battery is charged to. They may say in the instructions that they are not to be used for powering equipment including motors, or lighting. In fact the lighting caveat only applies to fluorescent tubes, and the motors caveat, to motors connected directly to the supply, which does not include modern telescope mounts, which use stepper or servo motors run off a controller, it being the controller that is actually powered from the DC supply.

All the 12 V equipment I have come across works correctly off these supplies. They are commonly available in capacities from about 3 to 10 A. The current capacities are quoted as two figures, a higher surge current and a lower continuous current. The total requirements of your equipment must not exceed the lower figure. Dew-heating strips on telescopes, particularly large ones, consume quite a few amps at 13.8 V, so if you have several of these, say, round the main telescope, round a guide scope, a finder and maybe a diagonal, it is wise to get the biggest supply you can. The small "wall-wart" or "fat-plug" type power supplies used with much domestic and computer equipment are of no use, they do not provide more than 0.5 A. An additional problem with these is that they cannot safely or easily be plugged into covered waterproof sockets. One needs power supplies fitted with a conventional compact plug and a lead.

Laptop computers use a variety of voltages, and these are very sensitive and must be exactly correct. They can be powered either using the manufacturer's adapter, from the mains, or using a third-party product that produces the correct voltage from a 12 or 13.8 V DC input. These are made mainly for running laptops off car batteries, and are non-critical as to the input voltage, but must be set correctly to provide the correct voltage for your particular computer model. They have substantial current draw from a 13.8 V supply, and the one I have gives an alarm sound if the supply voltage falls too low. If you go down this route (to save on mains connections), the supply to the computer needs to be factored in to your power-supply capacity calculation. Low-voltage telescope motor controllers require two to three clear amps by themselves, though the computerised handsets of GOTO models do not add much.

The final point on electricity is to re-iterate that, if in doubt about any installation, power supply or fitting, one should seek qualified, professional advice before proceeding.

Lighting

Lighting may be the astronomer's enemy, but, of course, we all need it. An observatory generally needs several levels of illumination, probably from different sources. A method of lighting the whole interior quite brightly, with white light, is very useful for working on equipment, setting-up, fiddling with small screws and recovering them when they get lost, and so on. This probably will be 230 V (Europe) or 110 V (North America) lighting, assuming mains has been installed, and could be fluorescent or incandescent light. The latter, being a warmer light, is perhaps more pleasant, and easier to recover from when adapting to the dark, and an additional advantage of incandescent bulbs may be the small amount of heat they generate, which can be used to reduce condensation. Again, fittings suitable for outdoor use must be used in an open observatory, and preferably in a dome as well. These will be better protected and insulated than standard light fittings. Again, if in doubt, consult a professional. A simple mains bulb screwed into a bayonet-type holder, fitted to the wall of an observatory, as one frequently sees, is not safe in the damp, and could well lead to a fatal shock. A correct, sealed, double-insulated external mains light fitting, as fitted in my observatory, is shown in Fig. 7.2. Also visible in this picture is an IP55-rated switch, mounted by the door, a handy hook for storing the padlock, and a digital thermometer.

Many favour red light in an observatory, though some go for green, which is almost the opposite, and seems to prove there can be no particularly clear-cut arguments about it. The idea of a dim red light is that it has minimal effect on the eye's dark adaptation. I have tried red lights in my observatories, but find that unless they are really rather bright, I can't usefully see anything, for example, to read charts or make notes. The argument advanced for green is more or less the converse, that the human eye is most sensitive in the green part of the spectrum, so the power can be minimised, which, it is also argued, helps one to retain one's dark sensitivity. Further research seems to be needed; I remain to be convinced about the benefits of these colours. If going for red lights, one will probably want to have brighter white lights as well.

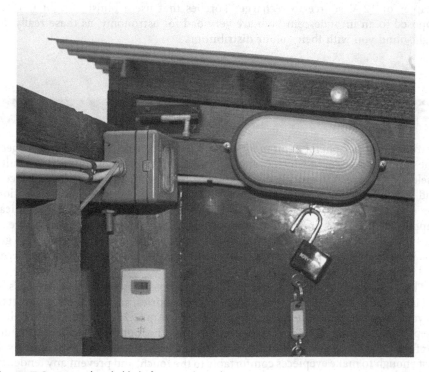

Figure 7.2. External sealed light fitting and switch.

One thing I have done is to install low-voltage bulbs, from an electronics supplier, in miniature screw-fitting holders, and to use a bulb rated at a higher voltage than the supply, for example, a 24 or 18 V bulb run off a 13.8 V supply. This gives a nice dim yellow light which I find quite good for reading and making notes by, particularly for lunar and planetary observing, where dark adaptation is not an issue. I also obtained a light on a gooseneck flexible stand, made for musicians to see their music, and fitted it to the top of the telescope pier. This is quite good, as it can be angled around to shed light when fitting and unfitting telescope accessories. I have a low-voltage lamp above my computer shelf, so I can see the keyboard, and had similar lamps on the analogue setting-circles on the mount, when I used a non-GOTO mount, so these could be read. With the declination circle travelling on the declination axis, which swung around the pier, I incorporated the wire to that bulb into the coiled, hanging, extensible cable that provided a different voltage to the declination motor. It is helpful to try not to bind a telescope up too much with lots of different leads to the various bits of it, though it is hard to avoid, particularly if your site is dew-prone and you need lots of dew heaters.

Going down another level in output and reliability are torches (flashlights). I find it best to have a few of these hung up around the observatory in fixed places, in addition to the mains and low-voltage lighting. You can buy red or red/white torches from astronomical suppliers, but I have not gone for these. Torches that use large, D-type cells are best, as they do not get exhausted so frequently. Rechargeable battery torches are also available, that plug into an adapter or 12 V source to recharge. Torches that use a bluish, bright LED as opposed to an incandescent bulb are very bad for astronomy, as these really do night-blind you with their colour distribution.

Storage of Equipment

Astronomical peripheral equipment is supplied, as, a rule, very badly packaged. The cardboard boxes in which eyepieces and the like are often supplied are quite useless. The boxes will quickly disintegrate, spreading fibres onto optical surfaces, and will provide no protection against dirt and damp. A better storage solution for eyepieces is the bolt-case. These can be bought from some astronomical suppliers. They are cylinders of plastic with a top and a bottom half, which thread into one another to close. They are not airtight, which means moisture can get out if an eyepiece is put in damp. Close-fitting caps over the ends of eyepieces and diagonals are less desirable.

Those good at making things can go a long way exercising their inventiveness in devising better storage solutions for optical accessories, such as shelves of metal, or varnished wood, with holes drilled to take 31.7 mm (1.25 in.) and 51 mm (2 in.) eyepiece barrels. An accessory cabinet would be another option, with compartments for filters. Such a cabinet, optimally, could be electrically heated to a very low level, just enough to make eyepieces comfortable to the touch, and prevent any tendency to dewing when they are taken out. Silica gel sachets are a good idea in cabinets and boxes used for storing optical accessories. These absorb moisture, and can be regenerated by placing for a few seconds in a microwave oven from time to time.

Cameras and electronic equipment, likewise, are best stored in dry, sealed boxes or cabinets. Books and charts present a particular problem, as they degrade so easily if allowed to get damp. The best charts for use in an observatory are those that are supplied plastic-laminated. The Webb Society in the UK do a very good set of such star charts. You may be able to get charts laminated, or to do it yourself, but sticky-backed plastic, I have found, is not particularly good as, if applied to only one side, it causes the paper or card to warp badly. It is very useful to have some reference books in an observatory, but, unless it is a dome, and additionally they are in a closed cabinet, they probably won't fare too well. They would be better kept in a "warm room", if you decide to incorporate one of those into your observatory plans.

Increasingly, I find I am relying on the computer for the functions that charts and books used to provide, in terms of obtaining data, ephemerides, star charts and finder fields, etc., used in the observatory. This is because a computer is much more damp-resistant than paper information sources, it takes up less space, it can be connected to the internet for the downloading of the latest data, and it makes many things easier to find. In remote-controlled observatory systems the computer can also control the telescope mount, cameras, focusing and filters, and feed programmed astronomical data such as ephemerides directly to the telescope. Automation of observatories will be discussed further in the next chapter. Although printed sources of information continue to be very useful, the relentless march of software does seem to be displacing them.

Cables

Another, related, trend is for the well-equipped observer to become bound up in more and more cables. Cameras, guide cameras, electric focusers, automatic filter wheels, the connections between these and computers and mountings, plus dew-heaters, fans, illuminated eyepieces, the list goes on, can result in a telescope becoming a spaghetti junction. Some of this can be mitigated by good observatory organisation. It is best to avoid having cables on the floor. Route them under the floor, which will be easy if you have taken the advice in earlier chapters to construct a timber floor with removable sections. It may be necessary to extend cables to do this, and fit new connectors, but anyone who practices a little with a soldering iron will be able to master this; it is not exactly electronic wizardry.

Some cables, such as low voltage power cables to servo-motor type devices such as simple electric focusers and motorised filter wheels, can easily be extended. This is not true for computer data cables, which should be bought in longer lengths, or extended using extension or repeater cables. USB connections can be extended using passive (un-powered) repeater cables, usually up to at least 10 m (30 ft), but some devices, particularly astronomical cameras, may not respond well to this. In such cases a powered USB hub may be necessary at the computer end of the sequence of repeaters to boost the signal. There seem to be no hard and fast rules with this, and it is necessary to experiment with your particular equipment. Webcams I have found to work well at the end of long repeater-cable runs.

Figure 7.3. Dew-heater cables tied to a mounting plate to minimise problems with snagging the mount.

Wires from the telescope tube itself should be tied to the mounting plate at the top of the dec. axis, to minimise the chances of them snagging anything or of getting pulled by the slewing of the telescope (Fig. 7.3). They, and wires originating at the mount, can be gathered together at the base of the mount and directed down the column. Simple plastic cable ties are the most useful accessories in this regard. Excess cable lengths should be tied up with these. Forward planning in construction might have provided some way of routing cables through the column and out under the floor. If you keep your working areas clear and unobstructed you will find it far easier to operate. Connections to portable devices, such as laptops, can be kept in one place by fixing them with cable ties or clips, so that when the device is removed, the plug and wire will not fall on the floor. By simple means the jungle of cables may be tamed, and so long as things in general are screwed rather than nailed down, changes to the system will be easy to make in the future.

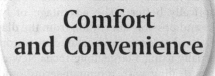

CHAPTER EIGHT

Comfort and Convenience

A certain amount of physical discomfort does go along with observational astronomy in general, the can be no denying that – an observatory will never be a beach, but discomfort should be minimised, both for maximising enjoyment of the hobby, and maximising alertness and productivity of the observer. I will say a few words here about things I have found that help in this regard.

Observing Positions, Chairs, Ladders, and Platforms

As we get older, most of us find that it is increasingly tiring to observe for long periods standing up. Standing is usually quite convenient for visual use of Newtonians, but not for refractors and Cassegrain-form telescopes, which are not normally mounted high enough for this, unless the telescope is being used at an unusually low altitude. Large, long-focus Newtonians, on the other hand, place the observing position very high, and need to be used with a platform, a ladder or a specially-made very high chair. Some kind of observing chair is a most desirable comfort for any observatory – or perhaps a selection of chairs at different heights, though this might cause too much clutter in a small observatory.

In the 19th and early 20th centuries, when the vogue was for long-focus refractors, a common kind of arrangement was an observing bench or couch, a bit like a weightlifting bench, with an adjustable sloping back, and an adjustable height. Elaborate examples, such as the one still used with the Northumberland telescope in Cambridge, England, took the form of a huge wooden arc on which the observing bench moved, so as to keep the observer's relationship to the telescope constant with changing eyepiece elevation. There are few contrivances like this left in use now, and the lying-down observing position seems to have gone out of fashion, most observers using refractors and Cassegrain-pattern telescopes preferring to sit in a normal chair and use a mirror or prism diagonal (the mirror type is optically better). The advantage of lying down is that it eliminates the light-loss and possible aberration from the diagonal, and it actually is comfortable, but it is hard to arrange in small observatories.

When it comes to visually using a large, long-focus Newtonian, or a very large, short focus one, we are dealing with mountaineering, and the risks that that entails. I would not put up with a large Newtonian on an equatorial mounting that did not have an easily-rotatable eyepiece position, or a choice of eyepiece positions. Without either of these, they are just impossibly awkward. The rotation can be accomplished either by rotating the whole tube in its cradles, which usually takes quite a lot of force, and therefore has to be done before an object is centred exactly, as the operation usually knocks the centring off, or by rotating only the top end of the telescope. Reflectors that have been constructed to allow this are rare, and it is sometimes claimed that they don't hold precise collimation when the end section is rotated. I haven't experienced them.

The best system short of this is to have, as I have with my 245 mm (9.5 in.) reflector (Fig. 8.1), fairly-loosely fitting cradles (which still hold the telescope firmly enough), a band of metal (known as a slip-band) clamped round the tube above the cradles, so that the telescope can be loose in the cradles but not fall through, and handles fitted round the tube to aid in turning it. In fact, handles are a tremendously useful, but utterly simple, addition to any moderate to large telescope that is not always used in GOTO mode. With a Cassegrain-pattern telescope they should be fitted near the rear cell, to aid moving the telescope by hand. They remove the temptation, particularly with inexperienced telescope-users who you might invite to use your equipment, of moving the telescope by yanking on the focuser.

With the classical large refractors the telescope handles took the form of a metal ring, like a ship's wheel, all round the back end of the telescope. I imitated this on my 254 mm (10 in.) Dall-Kirkham Cassegrain with a piece of 10 mm (0.4 in.) diameter copper tubing, such as is used in immersion heaters, which I bent to a circle of the correct diameter (the material is supplied already curved). I then soldered the ends together with a gas-canister-type plumbers' torch and mounted it on the telescope by holding it at intervals with copper plumbing p-clips screwed to the rear cell (Fig. 8.2).

I have found a cheap, plastic, self-assembly rotating office chair to be an invaluable addition to my observatory as an adjustable-height seat, to cope with most of the height variation of the eyepiece of a large Cassegrain telescope on a German mounting. It will not go high enough to cope with low-altitude observation, and there is an awkward range of intermediate positions for which the eyepiece is too high to use with the chair, but too low to use standing. Another cheap chair, a folding, aluminium and wood, high, bar-style one, copes with these intermediate positions. It is impossible to get a normal, adjustable

Figure 8.1. My Newtonian with a rotatable tube and handles. The small telescope stays fixed. Note also the sliding weight.

chair that is usable for all eyepiece heights encountered with this type of setup. I find the plastic office chair to be very useful as well, in that it is on castors, and so I can move from the observing position to the computer desk on it, without getting up. It is also small and light. I found steel office chairs, with or without arms, too big and heavy to be convenient in my small observatory.

I also have a small, aluminium, plastic-topped folding stool which I use in my run-off shed. It is stored in the shed when the shed is closed up. It is very light, but suffers from the disadvantage that, being metal, it conducts heat away, and the top, consisting of black vinyl plastic, radiates heat away, so if you are not sitting on it all the time, it becomes covered with condensation and has to be dried with a towel. Wooden or plastic chairs or those with fabric tops suffer much less from condensation. A wet bottom is not conducive to observing comfort.

These are ad-hoc, cheap solutions. Special, variable-height observing chairs are manufactured and sold by astronomical outlets. They are quite expensive. They are usually folding, for easy transport, and attempt to achieve a much larger height variation than can be had with ordinary adjustable chairs. Ideally, an observing chair should have a seat height range from about 30 cm (12 in.) to about 1 m (40 in.). I am afraid I simply haven't come across a specialised observing chair design that is completely successful (though I haven't tried them all), and I stick to my ad-hoc solutions. This is not to say that a successful portable observing chair is not possible. I don't know if it is.

Figure 8.2. The circular handle I made round my Cassegrain cell – Also showing home-made filter slider and ATK 1 HS camera.

One type is shown in Fig. 8.3. This is good in that it is made of wood, and so has little tendency to get dewed-over. However, the three bare, wooden slats that constitute the seat make it uncomfortable to sit on, and the seat itself is far too shallow for any normal human bottom. Were it deeper, the centre of gravity of the observer/chair system would probably fall too far forward, and both would fall over. The height is adjustable by fitting the seat to grooves on the front piece of the A-frame. However, it is restricted at the lower end of the range by the hinge, so it will not go as low as would be useful. Towards the top end of the range it becomes unbalanced and unstable-feeling, and there is nothing to lean one's back on, and, more importantly, nothing to put one's feet on, which are left dangling uncomfortably. Though the chair folds, it doesn't easily fold flat, as the seat will stick out unless removed from the frame completely. Overall, it is a poor design.

I have seen a picture of a chair manufactured in Canada, known as the Beer observing chair, which seems to avoid some of these faults. It is similar, but the top of the A frame is much higher and the seat slots in below it, so the hinge does not obstruct a large range of movement for the seat. It also has foot rests, and looks more stable. A not dissimilar A-frame design in metal and plastic can be bought, much more cheaply, from major stores, as an "ironing chair". These have been found useful by some astronomers, but they lack foot rests. As I say, I don't know if the perfect observing chair is achievable – the problem seems so far to have defeated human ingenuity – and the most comfortable, though not the most convenient, solution may be several normal chairs of various heights.

Figure 8.3. One type of adjustable observing chair that does not work well.

For visual use of a large Newtonian, another option is to place a high chair on a platform, but this is cumbersome. Some observers use step-ladders to observe standing up, which will be safe enough, provided one does not have to climb up too far, but tiring. It is fatiguing to have to stand on the narrow treads of ladders for very long. Easier to use are specially-constructed sets of wooden steps, equipped with a rail or post that is gripped for balance. These, however, will be bulky and heavy, and difficult to move around the observatory. Overall, the users of long-focus Newtonians do not have an easy time, and they also suffer from a much higher risk of damaging ancillary equipment, by dropping it from lofty heights.

Warm Rooms and Automation

Once consequence of the electronic revolution in amateur astronomy has been that it has become quite common for warms rooms, for telescope and imaging control, to be added to observatories. Of course, professionals have been using these for a long time. The warm room could be in the house; it could be, as I use, an outbuilding near the observatory, or in could be designed into the observatory itself. If the latter, it had best be separated from the observatory by a short, cold corridor to minimise thermal interference. It also should ideally be placed to the north of the telescope.

Personally, I would advocate the warm room being a separate building. However, people who have constructed run-off roof observatories with warm rooms as part of the same shed, on the north side, have reported that it does not cause any thermal problems, because, with the run-off roof open, the observatory is entirely at outside temperature. Certainly, the warm room should not be underneath the telescope, as I have seen in some observatories. A first-floor observatory always creates severe problems of telescope support and mechanical isolation. A pier rising to a first-floor level is never going to be vibration-free, if it be of reasonable dimensions. Having the control room underneath adds to this problem the problem of rising warm air.

For basic remote control of a telescope, the requirements are: firstly, a method of getting pictures from your telescope to a remote monitor or computer; secondly, a method of controlling the drives at both fast and slow slew rates remotely; and thirdly, a means of controlling focus.

The pictures can come from a video camera, a surveillance camera, a modified webcam, or a CCD camera. Popular video cameras for astronomy are made by Minitron and Watec. These use an S-video connection to a monitor, and such leads can be very long. To connect them to a computer, for recording images, an electronic box known as a frame grabber is required. Most other cameras used for astronomy now have a USB 1 or USB 2 connection, but a few use FireWire (IEEE 1394). In the past, parallel and serial connections were used. As mentioned before, USB connections can have problems over distances of more than a few metres or yards. A series of passive repeater cables might work, but, if not, active boosting will be required. Another, more complicated, possibility is to set up a computer network using ethernet connections, which can be extended a long way (hundreds of metres or yards) without trouble. In this scenario, one computer in the observatory connected to the camera, and possibly also to the telescope mounting, would communicate, over the network, with one in the control room.

The way in which remote telescope control can be implemented will depend on the type of mounting and drive motors. For older mounts, and altazimuth telescopes, a bespoke system will probably be required. The use of synchronous motor drives is not really a possibility, as they cannot be slewed fast enough to be of use for remote control purposes. All other types can be adapted. The telescope could be controlled from a specially-made control box, or interfaced with a computer. A non-GOTO stepper or servo drive system could just have its handset lead extended sufficiently far to make it "remote".

For remote operation, the telescope's pointing direction, obviously, needs to be known. Therefore shaft encoders need to be incorporated into the mount, or step-counting electronics need to be used with stepper or servo drives. Attaching encoders to Dobsonian mountings is relatively easy, because of their large bearings. For German equatorial mounts that have not been designed with this in mind, it is a non-trivial problem. The systems developed by American amateur Mel Bartels offer methods of upgrading to automatic pointing, and remote control capability, almost any type of home-built telescope or simple mounting, including Dobsonians and other altazimuths. In the UK, AWR Technology are specialists in providing bespoke drive and control systems for observatory telescopes.

Most commercially-made new mountings are more consistent in their operation and capabilities, and the setting-up of a remote control link is going to be far easier with an up-to-date mount, which has been designed to interface with a computer. The main issue for most is the very high cost of computerised mounts that are capable of taking observatory-size telescopes, and that are genuinely accurate enough in their operation to make remote-control a realistic option. The leading manufacturers of such mounts are the American companies Astro-Physics, Software Bisque, and Losmandy, and the Japanese company Takahashi. All these companies produce very high-quality, heavy duty observatory mounts integrated with the latest control technologies, which work with a wide range of proprietary and third-party software. While mountings further down in the price spectrum may claim remote-control capabilities, the extent to which these can be put into practice is, in my experience, limited.

Typically, modern mounts use an RS-232 serial connection to communicate with a computer, and to update their own internal software (firmware), the latest versions of which can be downloaded from the Internet. This is curious, as RS-232 is an old type of connection, and a not very fast one, which many PC manufacturers have now abandoned. It is necessary to buy another piece of hardware, a USB to serial port adaptor, to connect mounts to many modern computers. By this means, planetarium programs, showing the sky for your location in real time, and containing the co-ordinates of thousands of celestial objects, can send commands to the mount to slew to particular objects or particular co-ordinates in the sky. If using a USB to serial adaptor, a virtual COM port is created in the system software, which is allocated a number, and this number must be set to correspond in the planetarium, or telescope control, software, and the virtual port software (accessed on Windows computers though the Device Manager). This can be a source of problems, as the computer can change the numbering of these virtual ports arbitrarily, without telling the telescope control software. If you have several adapters, it can be difficult to determine which number corresponds to which function.

Clearly, this type of remote control will only be helpful if the slewing is accurate enough. The accuracy must be higher, in arc minutes, than the field of view of your camera, or you will be permanently lost. This is the main reason why a top-quality mount, with top-quality encoders and motors, and minimum flexure of both the mount, and the optical tube, together with excellent, rigid coupling between the two, is pretty much essential for remote telescope operation. A partial way around these rigorous requirements can be to have a camera on a smaller telescope on the same mounting, acting as a digital finder or guidescope, also connected to a computer, which gives a wider field of view than the main telescope/camera combination. This is a method I have used to facilitate remote planetary and lunar imaging with a non-GOTO mount, by setting the pointing approximately in the observatory, and then repairing to the warm room to make fine corrections using the view through the guidescope. However, this is only a very limited form of remote control.

RS-232 connections, using Category 5 cables, can be reliable over distances of 50 m (160 ft) or more, making practical the control of the telescope from the house. My warm room is only a few metres from my telescopes, however, and I have found that, in my case, an easier system is to use the USB connections, already set up between the telescopes and warm room to transmit images, for telescope control purposes as well. I use a USB hub at each telescope, with the serial to USB converter, for telescope control, connected into it alongside cameras, and then have these hubs connected together at a further hub, which sends all signals, through only one wire, to the control room. This system seems to work perfectly for controlling two telescopes, one long-exposure camera, and one webcam, simultaneously. Certainly, if one went too far in this direction, one could exceed the maximum USB 2 bandwidth. However, mount control signals require very little bandwidth; it is fast cameras (webcams) which need a lot.

My system is shown diagrammatically in Fig. 8.4. The reason there are two RS-232 connections from the first mount to the network is that the Astro-Physics GOTO mount I have allows this, and it is needed to control the mount with two pieces of software simultaneously, which can be desirable (a planetarium program

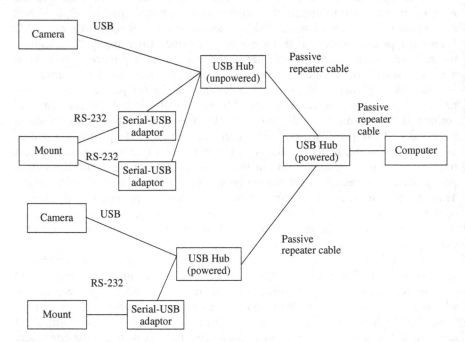

Figure 8.4. Schematic diagram of my observatory's USB data network, allowing control of two telescope/imaging systems by one computer.

and a tracking-correction program, for example). Most mounts, however, do not have this capability. It might be noted that Software Bisque's Paramount ME allows direct USB connections (so far as I know, the only mount to do this), thus eliminating the USB to serial converters, with their annoying software complication. The reason I have the second of the first-stage hubs powered, but not the first, is that I find this is necessary, as my second telescope is further from the second-stage hub than is the first, and is connected to it with two chained passive repeaters rather than one, with more signal loss.

I can place my laptop either in the observatory, connected directly to the powered hub, or in the warm room, connected via the repeater cable. The system thus allows flexibility, and only one data connection needs disconnecting and reconnecting to make the location change. Additionally, guide cameras can be connected to the hubs to provide serial autoguiding capability for the mounts (of which more later) without the need for any additional wires. This is subject to bandwidth restrictions, and software compatibility of multiple cameras connected to one computer.

If you are going to be a long way from your telescope while it is operating, you do need to be certain that various other things are not going to happen out there, unbeknownst to you. You need to be sure that the optics will not dew-up, so you need to have all the optics that can dew equipped with dew heaters, switched on. On all catadioptric telescopes you will also need a substantial dew shield. If you are using a dome, you need to know that the telescope is not staring at the dome rather than the slit. You need to know that the telescope will not collide with its mounting or with anything else in its movements, that it will always behave correctly, and that it will not get tangled up in its own wiring and break something. Ensuring that none of these bad eventualities ensue is the most difficult part of remote telescope control, particularly if you are completely isolated visually and aurally from the telescope. There are many possible mishaps or even calamities in remote telescope operation that it is difficult to predict, and for that reason, control from a nearby shed or semi-detached section of the observatory is likely to be an easier and safer option than retreat to the house. In any case, many people will wish to be secluded well away from family, TVs, radios, and other interruptions while observing.

One issue with remote operation of the observatory is that audible cues for how equipment is working are lost. One does not realise how important these are until the sounds are not present. One gets used to hearing how the drives should sound when they slew properly and unhindered, or how a focus motor sounds moving a small distance, and all these cues are incredibly useful. One idea of mine, which I have not tried, is to put a microphone, connected to an amplifier, on the telescope mount, connected to a speaker in the warm room. This might make the remote-control experience more "immediate", and could potentially prevent problems.

One important tip I have for remote-control imaging using a computer-driven German mount, which also will prevent problems, is to prevent unnecessary meridian reversal or normalisation – in fact to have reversal never happen, if possible. Most computer-driven GEMs will *insist* on reversing or normalising when slewing to an object on the "wrong" side of the meridian, which is just what you do not want when using a complicated system of telescopes and cameras with wiring attached to them. You do not want the wiring getting wrapped round the telescope and pier, possibly snagging, and jamming the mount or breaking something, which can occur even if the wires are well-organised, as I have recommended elsewhere, using cable-ties. In general, one will be working short distances either side of the meridian, and, provided one has checked that the required latitude of movement is available to the mount without the telescope striking the pier, there is no reason for the reversal, which, anyway, makes collimation, guidescope alignments, and slewing all less accurate.

Astro-Physics mounts and their associated software allow the meridian to be temporarily *offset*, so the telescope will behave as if objects past the meridian have not crossed it, or, with a negative offset, as if objects that have not passed the meridian have passed it. Some other computer-controlled mounts can have the meridian offset feature enabled through control software on a PC. The cheaper mountings on the market cannot, which severely limits their usefulness for remote operation (but, in general, their slewing is not accurate enough to allow it anyway). An exception to all this is the Paramount, which is designed in such a way as to maximise slewing and tracking accuracy, but to allow very little traverse on the "wrong" side of the meridian. The Paramount will need to be normalised a lot; however, it makes up for this by featuring through-the-mount wiring, which potentially can eliminate all problems of wires snagging on equipment.

Software and hardware does exist for controlling domes, and for co-ordinating them with telescope movements. The leader in this area is Software Bisque, which offers full options for co-ordinating their mountings with domes, with cameras, and with motorised filter-wheels for taking tri-colour images, though their software offerings *Orchestrate*, *CCDSoft*, *TPoint* and *AutomaDome*. There is a web group dedicated to sharing home-designed dome control systems. However, this level of sophistication is not essential for most observers engaged in typical observational programs still to benefit from the comfort of remote-control for much of the time.

For instance, a typical procedure of a CCD imager might be to slew his telescope to a bright star in the area of the sky in which he wishes to work, while in his dome, then move the dome by hand to the correct position, ensure that the camera, and possibly filter wheel, are in place and connected, and then retire to the warm room. In the warm room, the image can be focused on the camera chip and the star can be centred, and then, because the identity of the star is known, the control software can be synchronised exactly with the telescope's true pointing direction. The software can then be instructed to slew the telescope to the real target, say, a nearby faint galaxy, the camera exposure can be increased sufficiently to reveal the galaxy, further corrections can be made, and then the image capture can be begun with the capture software, all operations taking place on one computer. This procedure eliminates many of the possible errors in automatic telescope pointing, by restricting operations to a small area of the sky. Usually a long sequence of exposures will be taken, possibly through different filters, which will mean quite a time in the warm room before it is necessary to venture into the cold observatory again. If several targets are planned close together, this interval may be even longer.

Amateurs specialising in survey work may need complete automation for the high efficiencies they require, for example, for imaging a large number of galaxies each night to monitor for supernova explosions, or a large number of star fields to look for comets or minor planets or new variable stars. But most observers are in a more casual league, and a partial level of remote-control will be sufficient for them. Although, as I have shown, multiple telescope control and imaging operations can be performed through one laptop, having more than one computer, or maybe one computer with more than one monitor, in the control room would make things easier in terms of being able to see imaging and tracking program windows simultaneously, and not having to shuffle open program windows so much to see what is going on. Some amateurs have quite a few computers in their observatories, dedicated to specific functions.

Remote focusing is an essential, and is in fact very useful in the observatory as well, particularly for high-resolution planetary imaging, where the large image scale means it is hardly possible to touch the telescope at all, without an absolutely first-class mount, in order to avoid losing the image. Electric focusing units in some cases connect mechanically to the focusing knob shaft that is already part of the telescope, and in some cases are separate add-on units, screwed or push-fitted to the back of the telescope (Fig. 8.5). They are usually battery powered. (Consistent with my policy of battery-avoidance in my observatory, I have wired mine to work off the main observatory 13.8 V supply.)

The lead between the focuser and control box, for most electric focusers, is a simple analogue one, carrying a DC signal and terminating in 3.5 mm jack plugs, and can easily be extended a long way. Some more sophisticated focusers use RS-232 control, and these can be connected to a computer, either directly through RS-232 leads, or via USB. Some mounts (such as the Astro-Physics models) allow the DC control focusers to be connected to them, allowing mount control software to control the focus using these focusers as well. Otherwise, a separate interface is required for this.

Most focusers, whether motorised or not, suffer from a certain amount of both focusing slop and backlash. The former means that the image shifts slightly when the direction of focusing is changed, due to a slight change of optical alignment. This can be a nuisance in imaging, though for visual work it is not usually a worry. The latter means that the focus does not change immediately when the direction of travel is reversed (similar to backlash in mount drive worm and wheel systems). An advantage of the add-on electric focusers, such as the Crayford types, particularly for telescopes such as SCTs, which normally use primary mirror movement for focusing, which generates both image-shift and backlash, is that these focusing irregularities can be eliminated for imaging purposes. A disadvantage of them is that their range is small, and does not allow access to the full range of focus that is gained in the moving primary mirror system. Systems exist for motorising the primary-mirror mechanisms of SCTs, made by JMI and Technical Innovations (the Robo-Focus system), both US companies. For ideal focusing remote-control of an SCT, possibly, two focus motors would be required, one to adjust the primary mirror position, and one to make shift and backlash-free adjustments using a fine Crayford-type focuser.

Remote-controllable filter wheels are very expensive, for what they are. I don't have one, so I go back to the observatory to change filters. However, automatic filter wheels can be connected to the PC as well (via the ubiquitous RS-232 so beloved of telescope accessory makers) and controlled with software, or they

Figure 8.5. Two types of electric focuser: a Crayford-type screw-on micro-focuser for SCTs (top), and a simple motor box added to a standard Newtonian rack-and-pinion focuser (bottom).

could have their own simple control box. Making one cheaply should not be beyond the powers of someone with basic engineering skills.

With most commercial mountings, below the most expensive level, it is wise not to slew them at their maximum rates, to preserve the gearboxes, which in some cases only contain nylon gears. They are annoyingly expensive to have replaced or repaired after the warranty period has expired, and always inconvenient, as this normally means sending them back to the manufacturer, who is often on another continent. One buys more piece of mind with high-end equipment, as well as more accuracy and predictability.

The true accuracy of most mass-market GOTO systems is not much better than the diameter of the full Moon, 30 arc minutes, and therefore, with longer focal length instruments, or smaller detector chips, the focal length of the imaging system needs to be appropriately reduced to always be able to locate the desired object in the field. This is done by means of a focal reducer between the telescope and camera. This lens acts in the opposite way to a Barlow or Powermate lens, causing the light rays to converge more sharply onto the detector, simulating a short focal length telescope, and increasing the field that can be viewed. Focal reducers can, however, have the undesirable side-effects of introducing field-curvature (so stars at the edges of the frame are not sharp), and vignetting (where the edges of the field are dark, because they are not fully illuminated from the objective). Common f10 SCTs cannot have their f ratio reduced below 6.3 before these effects become apparent, with moderate-sized detector chips (8 mm, $1/3$ in. chips or larger).

Another element in many remote imaging setups will be a camera connected to a guidescope (similar to, or the same as, the finder camera mentioned above), which is then connected to the PC again, to allow autoguiding software to send guiding commands to the mount. These commands can reach the mount by one of two possible routes. In parallel or ST-4 type autoguiding, which is a possibility with most modern mounts, whether computerised or not, they are sent from the parallel or USB port on the computer, via an interface box, to the mount, which they reach through a cable terminated with a telephone or modem-like connector, known as an ST-4-type guide cable (after the original SBIG ST-4 autoguider). In serial autoguiding, which works with computerised mounts, the commands are sent from the serial or USB port of the computer directly to the mount controller, using the same route utilised by planetarium-software control. This is the simpler option, using fewer wires. In both cases, the autoguiding software detects slight drifts in the tracking, due to mechanical inaccuracies in the mount, polar misalignment, and atmospheric refraction, by accurately monitoring the position of a guide star on the detector of the guide camera. It then automatically sends corrections to the drive motors to keep the pointing sensibly on track. This is absolutely essential for very long exposure imaging (five minutes or more), even with the best mounts. It works best if the effective focal lengths of the main and guide telescopes are comparable.

Autoguiding can in some cases be done by the main imaging camera simultaneous with imaging, eliminating the guide scope and camera, either by dual-use of the imaging chip (the Starlight Xpress Star 2000 system), or by use of a smaller chip within the main camera (as used in the latest SBIG units). It can also be performed with a guidescope and a stand-alone camera and controller system in the observatory, like the early SBIG ST-4, STV, and their successors, which were developed originally to control telescopes for long exposure film photography, ending the wearisome drudgery, for the imager, of staring at a star on crosshairs for hours on end, making the corrections to the tracking by hand. Such stand-alone autoguider units are still very useful for autoguiding DSLR shots without need for a computer. For computer-based imaging operations, however, the other methods of autoguiding are more convenient.

A guide camera, or a guiding chip, differs from a long-exposure imaging camera in that it has to be able to read off its images quite rapidly, typically, once every few seconds, or more frequently, to allow the drives to be corrected quickly enough for image-drift not to become a problem at long focal lengths. It therefore requires an adequately bright guide star to work with. A guidescope is not essential for autoguiding, even without imaging camera-integrated guiding, as a guide camera can also be connected to an image-splitting unit on the main telescope, known as an off-axis or radial guider, which sends most of the light to the main camera, and a bit to the guide camera. This method, with the integrated camera methods, avoids the problem of slight changes of the guidescope pointing with respect to the main telescope, due to mechanical flexure. However, these methods run the risk of there being no adequately bright star for guiding present in the main field. A separate guidescope has the advantage that it can be angled around quite a wide area of sky in order to locate a suitable candidate.

CHAPTER NINE

Some Case Studies

In this chapter we will look at some particular, contrasted small observatories, including the practical experiences of those who created them, their motivation, methods, and how the results ultimately turned out. Other examples were featured in the two previous Springer books on this subject, in articles authored by their builders;[1] but these examples have not previously featured in a book.

Bob Garner's Observatory: CCD Imaging from a Converted London Garage

The details of Bob Garner's observatory will be found to contradict a great deal of the advice in the earlier chapters of this book. Bob has made the best of a remarkably unpromising location for astronomical observation, using much ingenuity, and frequently shoestring means. His observatory may well be the

---

[1] *Small Astronomical Observatories* and *More Small Astronomical Observatories*, Patrick Moore (ed). The first book is on a CD ROM sold with the second book.

world's worst-sited, but he has managed to turn some disadvantages to advantages, and achieved remarkable results in the field of deep-sky CCD imaging – something one certainly would not expect from a location like his.

Figure 9.1 shows the road on which Bob's house lies. This is the A40, a principal arterial road that runs west out of London. It is brilliantly lit with high-pressure sodium vapour lamps on 15 m (50 ft) columns (much higher than the house roofs), and traffic thunders past day and night. The house is semi-detached, of typical British inter-war architecture, with an old garage to the side of it and a small, narrow garden behind it. Crucially, for astronomy, the house is on the south side of the road.

Bob first placed his telescope in his garage about 1975. At that time, he was using a 250 mm (10 in.) Fullerscopes reflector, on a Fullerscopes Mk IV mount (a good commercial telescope and mounting for the time), with setting circles, synchronous motors on both axes, and a 150 mm (6 in.) reflector guidescope, for film astrophotography. Bob found this setup was too heavy to move. He solved the problem by removing a panel of the slightly-sloping asbestos cement roof of his garage, and replacing it with a plastic panel that could be lifted off. By this means he found he could gain access to a limited part of the southern sky. The rest of the fabric of the garage blocked out the light from the streetlights to the north, and this "window" offered a view of the meridian area. The great storm of October 1987, which damaged or destroyed many small English observatories, blew this plastic panel away, and Bob created a more permanent solution, which has remained more less unchanged since.

Bob cut out, with a handsaw, most of one half of the asbestos cement garage roof, leaving in place only narrow sections at the side, in the half that had covered

Figure 9.1. The trunk-road a few metres from Bob Garner's observatory.

the telescope.[2] Below these side-sections, he installed aluminium rails on steel brackets jutting out from the walls, running the whole length of the garage. These rails are on an approximately flat plane, whereas the roof of the garage slopes gradually towards the south end. (With a structure that had an apex roof, such a modification would be far harder.) The missing roof section was replaced with plywood sheet mounted on a steel frame, fitted with small fixed castors which run on the aluminium rail. When the observatory is not in use, the moving roof panel is held vertically against the remaining fixed asbestos sections by three wooden blocks on either side (Fig. 9.2). These blocks are graded in length so as to hold the section up against the fixed edge sections, at the general angle of slope of the roof. Thus the roof continues to drain rainwater to the south. The moving section is also secured to the wall at the south end by strong springs.

When the observatory is opened, the blocks holding the roof section are removed, and it falls onto the rails, whence it can be slid northwards under the other part of the garage roof. This arrangement allows Bob to observe from 6° to 85° N in declination, with about 3 hours of RA available. Most of the rest is blocked by the walls and fixed roof of the garage, apart from a little extra horizon

Figure 9.2. The rail and block arrangement holding up one corner of Bob's rolling roof, in the closed position.

---

[2]Note that cutting asbestos-containing material is today not recommended without protection against breathing in the dust, and that asbestos-containing material needs special disposal.

that can be gained by opening up the back door of the garage as well (Fig. 9.3). This might be thought far too restrictive, but if the rest of the garage were not there, the site would be flooded with streetlight. This arrangement creates a small window of darkness – the northern sky would be unusable in this location with any design of observatory. The way the telescope is placed in the observatory does not even allow an observer to get between it and the south wall, because observing is never carried out on that side. The telescope always remains on the east side of the mount, never being normalised (reversed). This preserves collimation and finder and guidescope alignments effectively.

The garage-observatory conversion probably cost less than £100 in 1987. At that time, Bob's interest lay in long-exposure imaging using hypersensitised, or "hypered" films, treated using forming gas, a mixture of hydrogen and nitrogen. He still has the gas cylinder nestling in a corner of the observatory, and the aluminium hypering chamber he made. There were many problems with this process. With the fickle British climate, a hypered film could be prepared and a few frames used, only for it not to be clear again for several weeks, after which time the hypering had worn off. Bob could observe little visually from his site (the naked-eye limiting magnitude is about 3.5), and at that time he was considering giving up astronomy. The CCD revolution came to his rescue.

In the early nineties, amateurs started experimenting with the tiny CCD chips of that time, initially only for guiding, and then for imaging. In 1994, a book was

Figure 9.3. The garage observatory fully open, with a little extra scope provided by the open door.

published by Richard Berry, Veikko Kanto, and John Munger called *The CCD Camera Cookbook*, describing how a cooled astronomical camera could be made using a Texas Instruments CCD chip, and a group of amateurs, including Bob, set out to build these cameras. Bob had no previous experience of electronics construction, yet he succeeded. Along the way he experimented with, and improved, the design of the Cookbook Camera in several ways. The camera used Peltier cooling, and water cooling to remove the heat transferred by the Peltier unit. The water was circulated using an aquarium pump. This arrangement achieved a cooling of 40°C below ambient temperature, better than most commercial units. The pipes for the water circulation at the telescope were awkward, however, and there was sometimes a problem of frost forming on the detector. The cost of making the Cookbook Camera was about £500, compared to a cost of £2,500 for the equivalent commercial camera of the day, the SBIG ST-6.

Since then, the cost of commercial CCD cameras has fallen considerably, whilst performance has improved drastically. Bob has recently been using the Starlight Xpress MX7C colour camera, and the Starlight Xpress MX7 monochrome camera with narrowband filters (which transmit the wavelengths of hydrogen alpha, oxygen III and sulphur II) mounted in a manual filter wheel to produce multi-spectral images of faint deep sky objects, such as the 14th magnitude planetary nebula PK164.8+31.1, known as the Headphones Nebula (Fig. 9.4). The cameras are used with a desktop computer, protected from dewing by a purpose-made cabinet on wheels (Fig. 9.5). This cabinet also incorporates, at the bottom, a home-made power supply, originally constructed to supply 12, 15 and 9.7 V to the Cookbook Camera. Other accessories Bob has constructed include a light-box for illuminating the telescope with diffuse light for the purpose of acquiring flat fields (which average out the response imperfections in CCDs). With the modern CCDs, Bob finds this is no longer necessary.

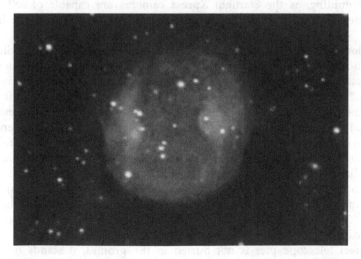

Figure 9.4. The Headphones Nebula, PK164.8+31.1, imaged by Bob using a 35 cm f4.6 reflector and a Starlight Xpress MX7 camera, total exposure 2.5 hours.

Figure 9.5. An open and shut case – Bob's computer cabinet on wheels.

The original telescope was changed in 2001 for a 350 mm (14 in.) f4.6 optical tube assembly, by Orion Optics of Crewe. Bob adapted the Fullerscopes mount to the larger telescope with new cradles, and he also created a door in the Orion Optics tube so the mirror can be closely capped, and hence better protected (Fig. 9.6). This picture also shows the heavy guidescope, mounted opposite the declination axis so as to preserve balance. The declination axis is loaded with 60 Kg (130 lb.) of counterweights. In fact the guidescope is no longer used for guiding, as the Starlight Xpress cameras are capable of self-guiding (using the Star2000 interface) using drift-monitoring signals from the imaging chip.

In order to use this system, Bob had to have the drive system of the Fullerscope mount modified. The original synchronous motors were replaced with a stepper motor system, by AWR Technology of Deal, Kent. One of these motors in shown in Fig. 9.7. Note the substantial metalwork required to brace the motor for the task of moving the big telescope. This system not only allows the telescope to be guided by the CCD cameras, it incorporates a GOTO function, whereby the motors will automatically slew the telescope to RA and dec. coordinates keyed into a handset. Bob finds that the telescope slews to within 1 arc minute of the target reliably, which is good enough to put the target on a CCD chip without having to search for it visually (which would be very difficult with his skies). In fact, most of the objects he images, he never sees visually. Contributors to this GOTO accuracy are probably the limited range of movement used for the telescope, and the fact that it is never normalised.

The steel telescope pier is not buried in the ground, it stands on feet on the concrete floor of the garage. Bob does get some interference from traffic

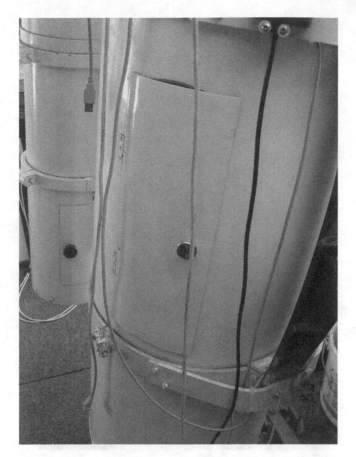

Figure 9.6. Home-built cradles and door in the tube of the 350 mm (14 in.) reflector.

vibration due to heavy vehicles going over the flyovers on the road junction near his house, and this factor, along also with light pollution, tends to restrict the CCD exposures he can use to about 3 minutes. He achieves longer effective exposures by stacking and averaging many such "short" exposures using the Astroart software package.

This observatory is close to the dwelling-house, but in this case, this is not a great disadvantage as observing is only done in the direction away from the houses. An observatory further down the garden would have suffered from the incursion of streetlight reflected off the next row of houses.

Bob admits that his observatory is somewhat crude, but it works, and the results he achieves, remarkable for a highly light-polluted site adjacent to large roads in the middle of a city, are anything but crude. It was, he says, a temporary solution that became permanent. His maxim is "If it works, why change it?" – and this dictum has served him well.

Figure 9.7. New stepper motor and bracket on the Fullerscopes mount.

Martin Mobberley's Plastic Shed Adaptation and His Telescope on Wheels

Martin Mobberley's name is one of the best-known in UK amateur astronomy. He was a pioneer of the use of video cameras for imaging the moon, and is also very well-known for his superb images of comets. Two of his earlier observatories, a run-off shed for a 49 cm (19.3 in.) Newtonian, and a run-off shed for a 30 cm (12 in.) Schmidt-Cassegrain, have featured in the previous "Small Observatory" volumes in this series.

Martin greatly prefers run-off sheds to other observatory types, for the all-sky experience they offer, and their environmental non-intrusiveness and tendency to not provoke problems with neighbours. In 2002 Martin acquired a new telescope:

a Celestron 14 SCT on a Software Bisque Paramount ME mounting. This is one of the sturdiest and most sophisticated mountings manufactured in the amateur range today. Martin wanted a highly-reliable imaging platform, mainly for comets, but also for supernova imaging and possible searches. The Paramount ME is a robotic mount with a control system and software that allows it to track the motion of any celestial object, even a comet that moves rapidly with respect to the stars. It can also be programmed to undertake unattended survey astronomy, its slewing control being integrated with camera control through the Bisque software. With automated astronomy in view, Martin preferred to do without the complications that a dome, needing automatic rotation, would involve. Additionally, he did not want the horizon obstruction that is inevitable in a run-off roof design.

Hence he embarked, with his father, a DIY expert, on his fifth run-off shed project. (The two others were both sheds for a 36 cm (14 in.) reflector, the first of which rotted away after 18 years.) With all this experience, he had discovered which designs did and did not work for run-off sheds. His early sheds had been too heavy and flexed too much on pushing. Sometimes, the effort required to use them had proved a real disincentive to observing. He also found that, with wood and roofing felt construction, they needed considerable maintenance.

A possible solution to these problems was presented by a plastic shed spotted by his father in a DIY superstore catalogue. The external dimensions were 2 m long by 2 m wide by 1.45 m high (78 × 78 × 57 in.). It had double doors at one end that were 1.3 m wide by 1.9 m high (53 × 75 in.). This would be adequate for the C-14/Paramount combination if the shed were aligned with the door parallel to the tube, which is about 1 m (3 ft) long with a motor focus and CCD camera attached. From the outer edge of the tube to the end of the counterweight shaft is 1.4 m (57 in.), so it would fit in with the shaft horizontal. The shed cost £249.

It was found that an observatory could be created from this shed with a minimum of alteration to its basic design. The shed was purchased as a kit, flat-packed. Firstly, the plastic floor, of cross-braced construction on the underside, needed to be adapted. Martin wanted to retain as much of the floor as possible, to strengthen the shed and to keep dirt out, but, obviously, there needed to be a gap for the telescope pier. The floor came as two equal panels, one for the door end and one for the other end. The door-end panel was cut into three sections with cuts at right-angles to the door, and the middle section was discarded. The other panel then had a curved recess cut into it. Extra bolts were added to strengthen the joints between the floor sections, and four pulley-blocks were bolted to the underside of the floor, almost directly under the walls, to act as wheels (Fig. 9.8). The shed was then assembled, which only took a few hours.

The site for the new telescope was to be that of the old 36 cm (14 in.) reflector, and the pier constructed for that telescope was re-used. This is a huge construction, consisting of two 1 m long by 53 cm diameter (40 × 21 in.) interlocking concrete drainage pipes. One is below ground and one above, and they were filled with concrete and capped with a metal plate when the pier was constructed more than 25 years ago. The area around was paved with slabs at that time.

Figure 9.8. Adaptation of the plastic shed floor, also showing one of the slabs and rails on the lawn.

With such a light structure as the plastic shed, wind was always going to be a concern, as well as security. Two metal hoops on plates were bolted to the underside of the new shed floor at the west (non-door) end to attach to hooks bolted to the concrete slabs, to keep the shed stationary when not in use. To further prevent movement, holes were drilled through the lower flanges of the walls and into the slabs. Cane bolts (draw-bolts with a large range of movement) were bolted to the inside of the sides of the shed so that they passed through these holes and locked into the slabs when dropped. Additionally, a wooden collar was cut to fit round the west side of the pier, and screwed to it with wall plugs. This overlaps the plastic shed floor, slightly above it, when the observatory is in the closed position, and so prevents any vertical movement. After some use, Martin felt the observatory was still too light and potentially moveable in the vertical plane, and added some small concrete slabs inside to weigh it down.

The rails came from the run-off shed of the old 14 in. reflector. They were originally acquired from a scrap dealer, and are an inverted T shape. Only the area immediately around the shed consisted of flagstones, so the ends of rails had to be on the lawn. Narrow concrete slabs were inlaid into the lawn, and the rails were screwed to these with wall plugs, through the flange of the T-shape. Wooden buffers were added at the western end to prevent the shed being accidentally pushed off the rails. Initially, the rails were only screwed down with a few screws. Then the shed was experimentally run on the rails. The separation of the rails was fine-tuned until the shed moved effortlessly, and the rails were finally secured, as close to level and parallel as could be achieved.

An issue Martin had encountered with his previous run-offs had been the ingress of dirt and animal-life (he talks of the terrifying giant spiders of Suffolk) through the open bases of the sheds. In this case, the remaining plastic floor should act as a partial barrier to these, and this barrier was augmented by the installation of a section of hardboard to fill the gap on the east side of the plinth when the observatory is closed (Fig. 9.9). This seems to have deterred the spiders.

The Paramount ME mount was purchased with the heavy duty base-plate, which was pre-drilled, and the metal plate already on top of the pier was re-drilled and tapped to correspond to this, allowing the Paramount to be bolted down. The mount weighs 30 kg (66 lb.), and the telescope 18 kg (40 lb.), the total weight of 75 kg (165 lb.), including counterweights and other hardware, being less than half the weight of the earlier telescope, of similar aperture, on the plinth: a striking aspect of the changes in telescope technology over the intervening period. The observatory and telescope was entirely put up by the two-man team of Martin and his father.

Part of the idea of this observatory, as mentioned, was that it should be operable remotely from the house, once the run-off shed is off. Thus there was the issue of providing cabling between the telescope plinth and the house. The cables required were a PC USB port to CCD camera cable, a PC serial port to mount cable, and the focuser control cable, as well as the mains supply. These were run in a flexible conduit hose under the lawn. To provide a USB extension over the 30 m (100 ft) distance required, Icron USB extension boxes were used. One of these boxes resides at the observatory end, and one at the house end, and

Figure 9.9. The shed in the closed position, showing sealing of the floor.

they are connected with Category 5 cable. This is probably a neater and more reliable method than using daisy-chained USB repeater cables. The Icron USB Ranger box at the observatory end has four output ports, which would allow multiple devices to connect with the computer indoors, if required. This system has worked well, as has the communication with the Paramount over an RS-232 (serial) connection. The mount is controlled using Bisque's *TheSky* software.

Remote focusing of the Celestron is performed with a JMI motorised focuser. This can be controlled in two ways. Using another Category 5 cable and RS-232 connection, it can be controlled from a PC – this function is also included in the Bisque software. Alternatively, the JMI digital handset can be brought indoors, and will work over the Category 5 cable. I have commented earlier on the fact that RS-232 is an old standard, "going out" on modern computers, and that the astronomical manufacturers' insistence on sticking to it is increasingly requiring astronomers to use USB to RS-232 adaptors, which add another layer of potential software/hardware incompatibility.

A final noteworthy point about Martin's C-14 setup is that he found a cheap way around the "dew-shield problem". This particularly afflicts catadioptric telescopes. Refractors, even cheap ones, are invariably supplied with reasonable dew-shields which limit condensation on the objective. Catadioptrics, even the most expensive ones, never are. This is totally incomprehensible, as it makes them unusable "out of the box" pretty much in any climate other than a desert one. In any normal location, dew will quickly form on the front corrector plate at night. It is rather as if automobile manufactures did not supply windscreen wipers as standard. Maybe SCTs are all manufactured by people living in deserts – but I don't think so.

Anyway, Martin's anti-dew system consists of a dew heater strip that wraps round the corrector end of the tube, a dew heater controller made by Astro Engineering, which takes a 12 V supply and modulates it to provide the minimum necessary power to the dew-strip, and a dew shield which was made from a flexible plastic offcut, spotted in a DIY warehouse, and acquired for a nominal sum. This offcut was rolled into a cylinder and reinforced with metal hose-clamps round the circumference, held together with nuts and bolts. These needed fine adjustment to get the diameter of the cylinder just right to be a friction fit on the C-14. The inside of the dew shield was then coated with matt black paint to reduce reflections. This saved the US$150 to $300 (plus UK import taxes) that a ready-made rigid dew shield for this telescope would have cost.

Since 2002, the C-14/Paramount/plastic shed observatory has been in frequent and successful use, principally for imaging comets and galaxies containing recent supernovae, with an SBIG ST9XE camera. However, more, recently, Martin has become very interested in planetary imaging using webcams, following the pioneering work on this by observers such as Don Parker and Damian Peach. They were using simple, cheap, non-astronomical cameras connected to computers, in conjunction with very long effective focal length telescopes, usually long-focus Newtonians and SCTs fitted with Barlow lenses, and image registering and stacking software, to produce remarkably detailed planetary images, surpassing anything that had been produced before using earth-bound telescopes. Many amateurs like Martin, who had previously tried to image planets using film, but become frustrated due to the limitations imposed by seeing, became enthused to try again using this new technology, and the discipline of planetary webcam imaging was born.[3]

The principal discovery made by these observers was that the traditional limitation of the resolution of large telescopes to perhaps the resolution of only a 15 cm (6 in.) telescope, on most nights, due to seeing, was overcome by the process of stacking and averaging very large numbers (thousands) of images taken in rapid succession, and that large amateur telescopes could produce images by this technique that equalled or even exceeded their classical theoretical resolution limit (the Airy Limit), if two conditions were met. These were that the surface accuracy of the mirror had to be excellent, of the order of one-eighth wavelength peak-to-valley, and that the collimation had to be very precise.

Though he already possessed a considerable stable of telescopes, Martin decided to buy one specifically for planetary imaging. His criteria were high optical precision, a fairly long focal length consistent with manageability and ease of use, a system that would cool down quickly to eliminate tube-currents and thermal distortion of the optics, and one that would retain its collimation well. It had been found that, although the high surface-brightness planets, Venus, Mars and Jupiter (plus the Sun and the Moon) could be effectively webcam-imaged using telescopes in the 15–20 cm (6–8 in.) bracket, a larger aperture was required for Saturn because of its low surface brightness. Consequently, Martin selected an Orion Optics of Crewe 25 cm (10 in.) f6.3 Newtonian on a Vixen Sphinx German equatorial mount for his planetary-imaging telescope.

Already having two full size run-off sheds in his garden, Martin decided that something low-profile would be in order to house this new addition. He reflected that he did not want his garden to look like a collection of chemical toilets. A paved area by the house was a possible observing site, with a good horizon from south-east to south-west. The obstruction of other directions by the house did not matter for planetary work. He decided to try the unorthodox system of having the shelter fixed, and the telescope moving on rails. This allowed a saving on materials: the house wall formed the back of the shelter, the patio formed the floor, and the shelter needed only two shallow side-walls, a sloping roof, and two wide doors (Fig. 9.10).

The only concern with this plan was how the telescope would fare if it was moved on rails. Would the polar alignment remain accurate, and would collimation be unaffected? The experience has shown these concerns to be unfounded, though this system would not be advisable if the telescope were to be used for long-exposure imaging of faint objects, when very precise polar alignment would be more of an issue. Planetary imaging runs last only a few minutes, typically (longer would be pointless because of the rotation of the planets), and even a rough polar alignment will hold them on the detector for this period.

The shelter was made 196 × 122 cm (78 × 48 in.), with a roof sloping from 130 cm (52 in.) height at the house to 117 cm (47 in.) at the doors. The basic framework was formed by 2 × 1 in. timber battens, screwed to the brick wall and to the paving slabs using wall-plugs. The walls and roof were formed of sheets

---

[3] For more on this subject, see the *Lunar and Planetary Webcam User's Guide* by Martin Mobberley, in this series (Springer, 2006).

Figure 9.10. Martin with his 25 cm planetary Newtonian on wheels, and its shelter.

of 9 mm (0.35 in.) thick plywood. These were simply screwed to the battens, and further battens were screwed on along the tops of the side panels, and joining the outer top ends of the side panels to support the roof above the door. The roof panel was covered with roofing felt, and the timbers were painted with green preserving wood-stain. The timbers in contact with the patio probably run some risk of rotting due to being in contact with a potentially saturated surface. In retrospect it might have been better to separate these from the floor with a damp course. The doors to the shed were made of more plywood, ledged and braced with timber. (Personally, I am not totally convinced about the use of plywood outdoors. Even the "external" grade tends to warp and delaminate with exposure. If the preservation is really diligent, I suppose it might last.)

Unusually low-profile rails were conceived to fit under the opening doors of the shed. These consisted of strips of white, flexible, plastic electrical conduit, having a channel-section shape and a total thickness of only 6 mm (0.25 in.). These were bought from a hardware chain, and were laid as a double thickness on the patio slabs, and screwed down using wall-plugs. The gaps between the patio slabs were first filled with concrete to smooth the telescope's path. Three rails were laid, for the running of fixed castors attached to the three feet of the telescope's steel pedestal. There is an end stop on the middle rail, but not on the others. This system provides a wide, unrestricted operating space around the telescope, with little possibility of tripping over the very low-profile rails.

The provision of mains electrical power to the facility, frequently a headache with constructions distant from the dwelling-house, was easy in this case. A hole was drilled in the house wall for the passage of a mains cable, and a waterproof double socket was installed inside the shelter just under the roof (Fig. 9.11). The overall cost of the shelter Martin estimates as under £150.

There were a few other refinements. Martin set up an artificial star (an Astro Engineering Picostar) on a pole at the end of the garden, 30 m away, for collimating the telescope. When seeing is too bad to collimate on a real star, Martin points the telescope at this device, which consists of a light shining through a 50 micron diameter fibre, and checks the collimation. The drawback of this method is that it is always possible, with any reflector or catadioptric scope, for the collimation to change when the scope is raised from the horizontal plane and directed to a planet. Martin also taped a thermometer probe to the side of his telescope mirror, and another to the outside of the tube, to monitor the difference in temperature between the mirror and the air. He reports that really good results are obtained only when the two are within a fraction of a degree Celsius of one another.

The only other addition to the telescope has been a JMI motorised focuser, essential for really precise on-screen focusing during imaging, when, using an effective focal length of 7.5 m (300 in.) and an imaging chip 4.5 mm (0.2 in.) across, touching the telescope at all would probably move the image off the chip.

Figure 9.11. Electrical supply in the shelter.

Having used some very large (by amateur standards) telescopes in the past, Martin has come to believe that user-friendliness and manageability are more important in getting observational results than telescope size. The biggest telescope and best-equipped observatory will achieve nothing if it is so daunting and complicated to use that the owner can never summon up the determination. This is the philosophy behind this minimalist observatory, which can be brought into action in seconds to take advantage of any fleeting observing opportunity.

Olly Penrice's Observing Retreat in the South of France

Olly Penrice and his wife Vic had long been lovers of France and all things French, and when they decided they had had enough of the weather and light pollution of England, they commenced their search for the ideal observatory site on the other side of the Channel. They had the idea of combining this, as a business, with a holiday retreat for astronomers and others.

In these matters, it is a reasonable plan for amateur astronomers to look at what the professionals may have discovered before them. In fact, French professional astronomers did a lot of research in the early 20th century to find the best site for an observatory within France, from the point of view of frequency of clear weather, good seeing conditions and transparent skies, and came up with the location of St Michel l'Observatoire, on a plateau 650 m (2,000 ft) up in the Alpes de Haute-Provence. This is where the Observatoire de Haute-Provence now stands. However, this site has since suffered from the spread of light pollution from the coast around Marseille. Olly and Vic decided that, though this is the right area, it was necessary to go further inland and find a more isolated spot. Also, they reasoned that, today, for visual astronomy, the best spot would probably not be the top of a high hill or plateau, but in a high depression or crater in the land, that would cut off light from human activities entirely.

They found this ideal location at *Les Granges*, an old farmhouse near the most appropriately-named village of Etoile Saint Cyrice, near Orpierre, Haute-Provence. Their site is at 900 m (3,000 ft), overlooking uninhabited mountains. No artificial light can be seen at all, but the mountains are low enough to give good horizons all round.

The observatory is equipped with a 0.5 m (20 in.) Dobsonian telescope for visual observing, and a 25 cm (10 in.) Meade LX200 SCT for visual observing and imaging, both housed in run-off sheds. (There is also a 15 cm (6 in.) Helios achromatic refractor on an undriven German mounting on a semi-permanent pier, and a 10 cm (4 in.) Televue Genesis apochromatic refractor on a portable Gibraltar mount.)

Olly has favoured run-off sheds for their compactness and ease of operation. He constructed these for both telescopes with square-section steel tubing frameworks. The Dob shed is clad in plywood (Fig. 9.12), and the SCT shed in shiplap timber, both varnished. The older of the sheds has been in operation now for three years without any problems, despite the quite severe climatic conditions that can be encountered in this location.

Figure 9.12. The 0.5 m (20 in.) Dobsonian emerging from its run-off shed.

Both sheds are now roofed with corrugated steel. Corrugated plastic was tried for the bigger shed, but it proved unsatisfactory in the heat of the Provençale summer. The roof of the longer Dob shed slopes at right-angles to the rails, while the squarer SCT shed has a roof that slopes to the back. The Dob shed has a single door, while the SCT shed has double doors, to save space on the somewhat narrow ledge of land on which is stands. The doors are framed in steel again. The Dob shed door was just covered in plywood, while, with the SCT shed doors, the framework was infilled with shiplap timber, leaving the steel edging to the doors visible, which makes an attractive effect. Figure. 9.13 shows the internal construction of the SCT shed. Note the diagonal bracing in all three dimensions for rigidity, and the shelf at the back for the dew-shield and other articles.

The steel tubing sections were bought from a local metal supplier, who just gave Olly a large radial arm grinder, and told him to cut off the lengths he wanted in the stockyard (unlikely to happen in Anglo-Saxon countries, with their litigious health-and-safety cultures). Olly bought an MIG (magnesium inert gas) welding kit for £150 ($300) and taught himself to use it in a couple of hours. The steelwork is finished with Hammerite paint (best applied, Olly has found, on a warm day). This has proved very durable.

The rails are U-section steel, 45 mm (1.7 in.) wide and 20 mm (0.8 in.) deep. These are used with hard nylon wheels, 80 mm (3.1 in.) diameter, mounted on fixed castors welded to the steel frames. Lengths of rail were joined together by welding to the rails overlapping side-plates at the joins. The rails were not bolted down at both ends, so as to allow for expansion in hot weather. Instead, flanges were added at the mid-points of the finished rails, and these were bolted to concrete blocks bedded in gravel forming the rail base, in the case of the Dobsonian shed, or to a concrete base, in the case of the SCT shed. The ends of the rails were left free, but constrained by short steel pins on either side of the rails, drilled into the concrete blocks in the case of the Dob shed, or into

Figure 9.13. The inside of the SCT run-off shed.

wooden sleepers bedded in gravel, beyond the limit of the concrete pad, in the case of the SCT shed. Thus the rails are kept parallel, but also free to expand and contract away from the central bolts. Before bolting, the rails were arranged on their bases and under the sheds to get the running correct. This stage of construction is shown in Fig. 9.14, with the framed but unclad Dob shed fitted round the telescope.

A slight drawback with the U-section rails bedded on gravel is that they can get stones trapped in the channel, so a brush is kept handy to clear them. Other than this, the sheds roll with very little friction, so that it is essential that their movement, when parking the shed, be stopped by something other than the telescope. In the case of the Dob shed, movement is stopped by the bar in the middle of the base of the framework running up against the concrete block on which the telescope stands, as shown in Fig. 9.14. Vertical movement of the shed, when parked, is prevented by a tongue on the chassis passing under a flange attached to the ground.

Figure 9.14. Arranging the rails for the framed, unclad Dobsonian run-off shed.

For the SCT shed, Olly devised a more elaborate arrangement which both stops the movement of the shed and holds it vertically (Fig. 9.15). A steel, inverted U, square-shaped stop is bolted to the ground. This engages with a latch device welded to the shed chassis, which clicks over the stop to prevent the shed from rolling back after it is parked. The latch is raised, when unparking, by pulling on a cable attached to a point near the door. Olly say that this simple device tends to make newcomers chuckle.

Additionally to these measures, both sheds are secured with guy chains and climbing-type carabiners (safety links with locking closures). Some "real weather", as Olly terms it, is encountered in these mountains. The chains do not seem to have done any work so far, but they ensure a good night's sleep when it is too cloudy to observe (Fig. 9.16). The rear guy chains do not need to be unclipped because they are long enough to let the shed roll back all the way, but are tight when the shed is in the parked position. The front ones do need to be unclipped every session.

Having used them for a few years, there are a few things that Olly would now do differently with the run-off sheds. A concrete base, as supplied to the SCT shed, should be sloped from the centre towards the edges slightly, so as to prevent water from running in and puddling. Also, there is merit in recessing the rails so there is no tendency to trip over them in the dark. Olly suggests this could be done by fixing wooden battens in place of the rails before the concrete is poured, then removing them when the concrete has set, creating a channel in which to lay the rails. Perhaps more easily, rails could be recessed by laying flagstones around them. Olly may do this in time, but he says it is not a big issue.

Figure 9.15. The latch on the SCT shed engaging with the stop on the ground to park.

Figure 9.16. The SCT shed parked and chained.

Once he had made the shed frameworks, Olly got slightly addicted to MIG welding, and was using his new-found skills to create numerous welded constructions of domestic and astronomical usefulness. A number of these are shown in Fig. 9.17, which shows the SCT shed run-off for use. The railing, obviously, prevents observers from tumbling over the precipice. The table has a varnished wooden top edged with wooden strips, to keep accessories from falling off, and it also has drainage holes, because it is outside permanently. The drainage holes are, of course, smaller than any of the items that might be placed on the table. There are two chairs: a variable-height observing chair, and one welded to the railing, which is just there to let one sit and admire the view. The seat base of the observing chair self-locks in any position under the weight of the observer, but there is also a quick-release strap connecting the top horizontal bar of the seat with the seat base, which is pulled tight at the chosen height, to stop the seat suddenly sliding down the rails (Fig. 9.18).

Also seen in the pictures is the welded tripod-pier Olly made for the Meade SCT. It has brackets to bolt it to the ground, and height adjusters at each point of contact with the ground, which can be adjusted to keep the pier vertical with any settlement of the concrete base. The tubes of the pier are sand-filled, for vibration damping.

A telescope like the 0.5 m (20 in.) Dobsonian presents its own problems of use. Climbing a stepladder in the dark, then leaning off it to peer into a telescope, which you are simultaneously trying to steer with both hands, is not a recipe for comfort or safety. Olly has addressed the problem, with the aid of his trusty MIG welder, by creating a wheeled observing platform, used in combination

Figure 9.17. The SCT shed open, with welded steel furnishings.

Figure 9.18. The variable-height observing chair and the Meade SCT.

with another variable-height seat, shown, in use, in Fig. 9.19. The platform has only two rubber-tyre wheels. The other two corners are the ends of square-section members, which sit in the gravel, so the platform cannot roll on its own. The platform is moved in wheelbarrow-mode, using the handle seen half-way up on the right. It is very stable in use, and cannot topple over.

The platform and chair combination allows four basic observing options: the seat can be used on the ground, the observer can stand on the ground, the seat can be used on the platform, or the observer can stand on the platform. These cover all needs with the 0.5 m Dob. On the platform, the observer is safe behind the handrail, comfortable, and well-braced, for two-handed guiding. The size of the platform allows the observer to spend long periods at the eyepiece without having to move it. As the object moves, the observer just moves along the platform. In addition, an observer and another person supervising him or her can work together in comfort on the platform. A plastic storage tub and a clipboard have been incorporated into the platform, sitting on the rails at the top, to complete the convenience of the package. The platform has greatly increased the user-friendliness of the telescope, and visiting astronomers, having experienced it, have gone home saying, "Our club needs one of these!"

Figure 9.19. The observing platform and chair combination in use on the Dobsonian.

The small bicycle-seat-style chair was not home-made; it came from Ikea, the furniture retailer. It is less comfortable than the larger seat, but gets more use because it is so convenient and compact. It is now discontinued, but Olly suggests one could be made from an old bike, using, of course, the welder. With respect to bikes, Olly also knows what he is doing, as cycling is his other passion, and the other reason he came to these mountains. This is classic territory for cycling, with the fearsome Mount Ventoux, of *Tour de France* fame, within sight, and to which summit Olly makes an annual pilgrimage. He and Vic market *Les Granges* as a centre for cycling as well as observing holidays.

An interesting feature of the atmospheric conditions encountered here is the temperature inversion, common to other mountain-top and island sites that are the locations for top professional observatories. An inversion in the atmosphere, where warmer air overlies colder, generally means excellent seeing. So normal is the inversion here that an old villager, to whom Olly talked, would not believe that the general rule is that temperature falls with increasing altitude. The local believed that it always rose, as, naturally, it should – after all, if you go up, you are getting closer to the Sun!

Quite differently to most astronomy holiday bases, the philosophy at *Les Granges* is to let visitors to get on with it themselves, once they have been sufficiently familiarised with the equipment. This means unlimited observing time, when the weather is suitable, which it is most of the time. Olly and Vic have managed to find an optimal location for astronomy in a reasonably accessible location in Europe, and they would love to see you there some time.

My Observatory: A Combined Run-Off Roof and Run-Off Shed Construction

My observatory, which has already been referred to a few times in this book, may be found at the far end of a suburban garden in Edgware, Middlesex, towards the north-western edge of Greater London. It was originally conceived to house a 20 year old 15.9 cm (6.25 in.) f8 Newtonian, and developed subsequently to house various larger telescopes and an increasing selection of imaging equipment. My speciality has always been the moon and planets, drawing, and now imaging them, although I also do general observing and have forays into imaging the Sun, comets and deep-sky objects. In addition, the observatory often hosts gatherings of local astronomers.

The basis of the observatory was an existing patio at the bottom of a 30 m (100 ft) long garden. The curious way in which the 1930s street plan was developed created a junction of individual wedge-shaped gardens just beyond this point, which are difficult to cultivate, and hence form a "wild" area quite distant from houses and streetlights, despite being in the middle of a town. This is quite good for astronomy. There is much vegetation, which tends to stabilise the seeing, and trees and hedges block out most light from house windows. The limiting naked-eye magnitude is 4.5 on a good night, which of course is a long way from a dark sky, but good for London. The northern direction has a high horizon due to trees, and the house blocks the eastern horizon, but most of the horizon towards the south and west is very good.

Previous owners had used the patio for barbecues, and the brick barbecue is still standing against one wall of the run-off roof observatory, constraining it, though it is not used. The patio was only a superficial construction of slabs on sand, stretching along the width of the garden from north-west to south-east, this alignment having been imposed on the area 2,000 years ago by the Roman Empire, who built the main road from London to North Wales, Watling Street, on this orientation, straight as a die.

The shape of the space and the sky visibility from various points suggested a run-off roof observatory located at the north-west end of the patio, with the roof running off along the patio to the south-east. I had built an opening-roof observatory before, and felt comfortable with the principle and, while I could see the attractions of a dome, I concluded that my skills were probably not sufficient to produce a satisfactory one.

I opted for low, pent-roof structure, with the telescope mounted quite high, so as to obscure the available horizons minimally with the shed walls. This meant producing a structure not high enough for an average adult to stand in comfortably when the roof is closed, and also created a necessity for the telescope to be carefully stowed with the tube horizontal and on the meridian, to allow the roof to close (Fig. 9.20). The observatory does have an excellent, usable low western horizon which has proved particularly useful for catching Venus at eastern elongations at southerly declinations, and enabled me to image it when

Figure 9.20. Telescopes under the run-off roof stowed horizontally on the meridian.

no-one else in England apparently was doing so, but the downside has been that I often have wished it were easier to work in the observatory on rainy days when the roof is on, without having to stoop. Such compromises are intrinsic to the run-off roof idea.

I wanted a strong observatory, as the previous one had been too flimsy, and a simple, reliable one, so I rejected the idea of an opening wall-section. In fact, the orientation of the patio site argued against this, as the meridian is in line with a corner, so there would have to have been two opening sides. The plan of the observatory site was given in Fig. 2.4.

There was some artificial light from the street direction, to the north-east, so I chose that to be the high wall side, with the door in it, which also made sense as that put the door on an existing path to the patio. The low horizon in the south-west was then catered for by having the lowest wall on that side. Wall obstruction to the north was not an issue owing to the trees. Since the only direction in which there was room to roll the roof was along the length of the patio to the south-east, all this meant having the roof roll at right-angles to its slope, which is slightly unconventional.

While I could have made the top section of the shed movable, including part of the high wall, so that both rails were on the same level, this would have made the door rather low, assuming the lower part of the shed had a continuous box-frame, running over the door. Hence I adopted the design I did, of rails at two different levels, held up at their far ends by brick piers of different heights, and only the roof movable, with no moving wall sections. This required some careful carpentry in making the roof so that the running members of it are vertical and horizontal, and the weight is borne vertically by the wheels, while there is a drainage slope on the roof of 11°, which is more than sufficient (Fig. 9.21). This design has been successful, in that the roof is light and easy to move by hand on the wooden rails, and it runs correctly even if it is pushed asymmetrically (which it needs to be, working from inside, with the telescope in the way), and the fixed part of the shed is very strong.

I started initially with the intention to make the observatory 2.4 m (8 ft) square. This was partly because this is a standard timber length, so there should be minimal wastage. However, this became compromised by the fact that the roofing sheets were also supplied as 2.4 m lengths. (I have since discovered that you can buy them longer.) Since I was intending to use only one length of roof sheet for the slope of the roof, and since that length had to be, by trigonometry, slightly longer than the depth of the observatory, and additionally, since an overhang of the roof is required, particularly on the draining side, to avoid water dripping off too close to the walls and floor and splashing back, this meant that the depth of the observatory had to be reduced. Its internal floor area ended up being 2.3 × 2.15 m (90 × 84 in.).

The observatory was constructed of relatively cheap timber – sawn and treated external-grade timber for the most part, with feather-edge timber cladding. The floor was initially laid down as six 3 × 2 in. (nominal size) battens, with gravel-boards (rough planks sold for fencing) screwed down as floorboards. An area of boards in the centre of the observatory, around the brick telescope pier, was

Figure 9.21. Detail of the run-off roof and rail construction, with one of the securing bolts.

designed to be removable. The first version of the telescope pier, built to support the 6.25 in. reflector, was only one brick stretcher (the long-side of a brick) square in cross section, which was completely inadequate.

The walls were then each assembled individually, as frameworks of 2 × 2 in. timber, with vertical reinforcement at an interval of about 40 cm (16 in.) – this was very conservative – the timbers jointed and screwed together. The walls with sloping tops had to have the lengths of all the verticals carefully calculated, and the joints cut to the 11° angle (Fig. 9.22). The featheredge cladding boards were then cut to length, and those under the sloping eaves were cut to the 11° angle, which made for some rather narrow, pointed laths of wood. The featheredge timbers were nailed to the frames to make the finished walls, which were then coated with wood preservative (creosote substitute) on the outside, before assembly of the shed (Fig. 9.22). The whole thing was a one-man job, though the wall panels were almost too heavy to move single-handed, and the roof required a lot of ingenuity to get it on without any help.

The walls were bolted to the floor with three coach-bolts each at the font and back. One long screw for each of the side walls, in the middle, sufficed to secure them to the floor, with four screws holding adjacent walls together at the corners. The number of screws and bolts was minimised to make the whole thing easily dis-assembleable, if required, at a future date (as in fact it was).

The rails went on next. These were two pieces of 2 × 4 in. timber 4.8 m (16 ft) long. I decided single lengths of timber would be best for the rails, to avoid

Figure 9.22. Shed partially assembled, showing unclad wall-frames, and 6.25 in. telescope experimentally in place on the first, unsuccessful, pier.

junctions in which the wheels might stick. It was more difficult to obtain these long timbers than the rest of the materials: DIY stores did not stock them, so a specialist timber-merchant was the answer. The rails were bolted to the walls using three large coach-bolts each. At the brick pier ends they are not attached, they just rest on the piers, allowing scope for expansion and contraction with the weather. The rails create an overhang at the tops of the walls of 5 cm (2 in.), duplicated on the sloping-topped walls by pieces of the same 2 × 4 timber, ends cut to the 11° angle. It will be noted that the walls, finished in this way, do not meet at equal heights at the corners. The front and back are lower than the sides, and the roof frame was made with unequal timbers that key into this, so that the roof can only move along the rails, not sideways. To make doubly sure of this, the rails are also edged with 1 in. square-section moulding timber outside of where the wheels run, nailed in place.

The roof was constructed of a framework of 2 × 2 in. timber jointed and screwed again, except for the timbers carrying the wheels, which are 2 × 4 in. with the long side vertical (Fig. 9.23). This was covered with "mini-profile" corrugated plastic sheeting. This is a bluish, translucent material, and is quite difficult to deal with. It requires eaves-fillers between it and the battens – expanded polystyrene strips cut to a matching profile. I found the best solution with these was to glue them in place on the timbers, then place the sheeting on them, with the required overlaps of at least two ridges, adjust the fit of the eaves-fillers while the glue was still flexible, and then start nailing down the sheeting, through the eaves-fillers, with the special spiral nails and curved plastic washers. This process often cracked the sheeting locally, and I waterproofed the small cracks with Evostik glue. I have since found it a better plan to drill the sheeting.

It is quite difficult to find suitable wheels for observatory projects. I found 7.5 cm (3 in.) plastic ones with rubber tyres, and no axle, just a hole in the middle, in a DIY store. (These later became unavailable.) I placed these on the ends of M6 coach bolts passing through holes in the 2 × 4 in. beams forming the front and back of the roof, and secured them with a lock-nut (two nuts tightened against one another). They were set up quite tight initially to allow them to wear-in. The inside ends of the coach bolts, with the heads, had to be slightly recessed into the 2 × 4 timber to prevent them from obstructing the roof movement. There are four wheels per side.

Figure 9.23. Constructing the roof framework.

The wheels were fitted after the roof was in place. It was placed on the shed by creating a 45° ramp out of an aluminium ladder, and sliding it up. Help would have made this process much easier, but I did prove that an observatory could be constructed by one person, working alone. I also constructed it without any power tools apart from a drill, though, in retrospect, this was foolishly puritanical. One small mistake I made with the roof, already referred to in Chapter 5, was to construct the cross-halving joints in the framework shown in Fig. 9.23 the wrong way round, with the continuous surfaces of the front-to-back members (at right angles to the rails) at the top, which allowed the roof to bow between the rails. I had to mitigate this error by bolting reinforcing battens to the sloping sides of the roof. These reduced the tendency to sag to an acceptable level.

The roof was fitted with tower bolts (draw-bolts) at all four corners inside (shown in Fig. 9.21), engaging with holes drilled in the 4 × 2 in. roof beams. The operating sequence is to first unlock the door, then go in, withdraw the bolts at the corners, and then push the roof back from inside just by pushing on the roof battens. To close up, there is a handle on the outside edge of the roof abutting the shed, which is pulled on from inside to close the roof (while ducking down).

I found, to my cost, that the bolting of the roof really is most important in windy weather, if the door is left open. If the door is shut there is no problem, but on one occasion I left the door open and the roof unbolted, and went into the control shed next door. A freak gust of wind then hit from the north-east, went in the open door, and blew the roof off. It landed on the hedge behind, and some of the plastic panels were broken and needed replacing. The lesson is to never underestimate freak winds. In this case the roof was lifted by a pressure difference that arose when the gust entered the door and had nowhere else to go. When the roof has been in the open position, it has never been lifted by the wind (but I would not try leaving it open in a severe gale).

The door was constructed of gravel-boards, but a higher quality of timber would have been preferable, as it warped, and needed to be heavily reinforced later. It is fitted with a hasp and padlock, handles inside and out, and ball catches at the top and bottom, which keep it closed when unlocked. It is hung using three small T-hinges. The hinges screw to the shed frame through the cladding, and the slope of the featheredge surfaces was evened out under the hinges using small inverted pieces of featheredge. The door is stopped on the inside by a sub-frame of 1 in. square timber moulding, nailed to the door frame. The corners of the building were finished in the traditional manner for sheds, with rectangular timber mouldings, nailed to the frame, which give a better finish than the ends of the featheredge.

The observatory was carpeted inside with dark blue, 12 in. square, carpet tiles, which got thrown out of my father's house when he was doing some redecorating. More of these tiles were used in the warm room shed. They were stuck to the floor timbers with glue. These not only formed a much softer surface on which to drop delicate equipment, but also darkened the inside of the observatory, to help with dark adaptation, and made it easier to glide around on a small swivel chair with plastic wheels, which I use as an observing chair, and on which I propel myself around the shed. If desired, I can also fairly comfortably lie on the floor to use the big Cassegrain telescope without a diagonal.

The insides of the walls were initially left in their "natural" state, consisting of the interior of the featheredge boarding, but I found this could get slightly damp when rain was driven into it by wind over a long period, and also it was rather a bright surface, affecting dark adaptation (to the extent that this is possible at all with the skyglow experienced at this site). I decided to line the observatory with hardboard panels, which I cut to fit in the spaces between the wall frame timbers, so that the hardboard would contact the featheredge and not reduce the available space in the observatory at all, as an extra skin within the wall frames would have done. Hardboard is cheap and very easily cut, but it has no resistance to damp whatever, so I double-coated it on both sides with primer and undercoat, and applied external gloss paint to the visible surfaces. With the large number of small panels, this was a long job. Eventually the panels were fitted in, and held in place by small pieces of wood stapled to the framing timbers.

This thin, double-skinned wall construction does not retain heat in any noticeable way, which is good. The panels were painted dark blue to match the carpet and the main telescope. I think dark blue is a more attractive colour than black, but it is almost as effective at darkening the interior at night. The wall joists I coated with a dark blue water-based preservative made for use on un-planed timber.

Some shelves were fitted in the north-west low corner for eyepieces, filters, diagonals, and adaptors (Fig. 9.24). Most of the eyepieces are kept in bolt-cases, and the filters in boxes. I have attached numerous screw-in hooks to the shed frame inside, to give definite locations for all the various minor articles that need to be stored where they can easily be got hold of – torches (flashlights), clipboard holding paper and pencil, leads, and tools. I find it very useful to have a selection of tools including screwdrivers, spanners and Allen keys (wrenches) always in the observatory.

A hinged, quarter-circular shelf was created in the north-east corner, to accommodate a laptop computer (Fig. 9.25). This shelf can be lowered out of the way to maximise space for groups of people. The hinges are at the back, and it is held up at its front corner by a small draw-bolt that secures it to the shed frame. From this corner, extension USB, serial and DC focuser cables run across the gap between this shed and the warm control shed, via a piece of plastic waste water pipe. The laptop can either be used in this position, or in the warm shed. Since all the data can pass through USB, which is routed into one hub, located at the back of this shelf, the transfer can be done by disconnecting and reconnecting only one cable. All the cables from the telescope pier are routed to this corner under the floorboards. Easily-removable floorboards are a great asset.

Under the computer shelf are placed the power supplies. There is a dedicated 13.8 V, 7A supply for the mount, another, identical, one for the dew heaters, low voltage lighting, focusing motors, fans and CCD camera, a 15 V supply for the laptop, and a variable 3–4.5 V supply for a camera Peltier cooler. There is also a small dehumidifier, with its own power supply, visible in Fig. 9.25. Because this is left running when the shed is closed, it is switched from the mains entirely separately from everything else. There is thus reduced chance of leaving anything else running when the observatory is shut. This is most important for the mount, which could cause a calamity if left running. As an extra measure, the supply to the mount is always broken with *two* switches, one on the pier, on the 13.8 V

Figure 9.24. Shelves for optical accessories.

line, and one the 230 V supply. The weatherproof mains sockets are located under the computer shelf, switched by weatherproof switches, placed above the shelf. Power to the other telescope shed can also be switched from here, and power to both sheds can be switched from the warm room. Mains lighting is switched separately, using more weatherproof switches near the door. There are two bulkhead lights at the two high corners of the observatory, to provide general bright lighting when necessary.

As mentioned, the initial narrow brick pier supporting a 156 mm (6.25 in.) Newtonian was far too feeble, and, when a larger telescope was installed, this was built round, or encased, with another layer of brick, to create a brick pier two stretchers (brick long sides) square. This was strong enough, but rather bulky, and got in the way both of observers, and the telescope, in certain positions. Eventually it was replaced with a cylindrical concrete and steel pier, as already mentioned in Chapter 5. The brick pier had simply been built on the patio slabs, which had no real foundation. The new pier was given a concrete foundation

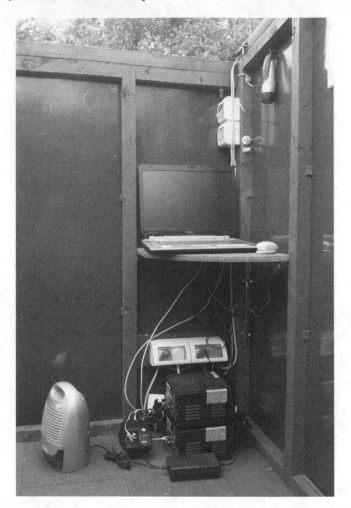

Figure 9.25. Computer shelf, power supplies, mains switches, and dehumidifier.

about 1.2 m (4 ft) deep, with embedded steel pipes driven into the London clay subsoil about 2 m (6 ft). To demolish the old pier and excavate for the new one, it was necessary to completely take down the shed. Because of the modular construction, with minimal fixings, this was accomplished by a friend and myself in one afternoon. After the new pier had been cast, the shed was re-erected in another afternoon.

The new pier is fitted with a square shelf of varnished plywood at the top which has many useful features. One is a control box from which power to the mount, cameras, dew heaters, fans and low-voltage lamps can be switched. Attached to this box is also a 12 V lamp on a gooseneck stand. Another feature of the shelf is a selection of 51 mm (2 in.) and 31 mm (1.25 in.) holes drilled with a holesaw, to fit eyepieces, adaptors, and cameras. Also attached here are hooks for hanging mount and focus hand-controllers, a bank of sockets for supplying 13.8 V DC to the dew heaters, and another USB hub. There is now a lot of wiring on the pier, mostly associated with imaging equipment, which is held in check with cable ties so far as possible (Fig. 9.26).

The observatory now houses a 254 mm (10 in.) Dall-Kirkham Cassegrain telescope, fitted with a 90 mm refractor guidescope. This combination has recently been re-mounted on an American computerised mounting, the Astro-Physics 900, having previously been on a 35-year old, massive, synchronous motor-driven, cast-iron German mounting by the makers of the Cassegrain, Astronomical Equipment of Harpenden, England (Fig. 9.27).

The Cassegrain is used principally for imaging the moon and the planets. This unusual telescope has been described elsewhere.[4] Suffice it to say here that it gives an exceptionally long effective focal length for an amateur telescope, of the order of 5 m (200 in.) and above, well-suited to observing and imaging detail on the moon and planets, and to double-star work, but not well-suited to more extended

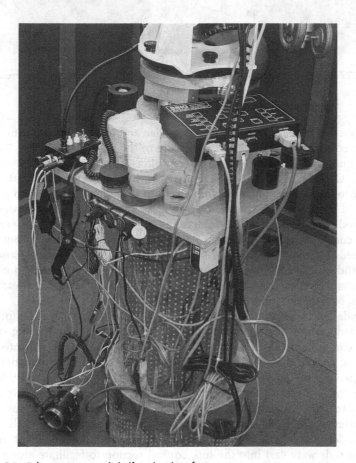

Figure 9.26. Telescope pier and shelf, with a lot of wires.

[4]Journal of the British Astronomical Association 2007, Vol 117 No. 2, p 85.

Figure 9.27. General view of the run-off roof shed, with the Cassegrain on its new mounting.

objects, such as comets, open clusters, nebulae and galaxies, which cannot really be observed with this telescope at all, owing to the large image-scale.

This limitation has prompted two additions to the observatory. One is a Celestron C-11 28 cm SCT, which happens to have a very similar outside tube diameter to the Cassegrain, and can be used in the same tube rings, to provide a wider-field telescope with the same viewing arrangement. This is, however, still only a moderate focal length instrument, at f10, useful mainly for observing and imaging compact deep-sky objects such as galaxies and planetary nebulae. I felt I also needed a short telescope of substantial light-grasp to observe and image comets and other very extended objects. Consequently it occurred to me to set up a permanent telescope of very low focal ratio at the vacant south-east end of the patio, beyond the limit of the run-off roof travel.

The wide-field telescope is a 245 mm (9.5 in.) f4.8 Newtonian reflector, home-constructed around mirrors by Orion Optics of Crewe, England, using variously sourced components and home-made ones. This was initially mounted on an aluminium tripod, and then on a concrete pier of similar construction to the one supporting the 10 in. Cassegrain, but narrower and lower (Fig. 9.28). This concrete pier was cast in the too-flimsy aluminium tube that originally held the optics, apart from the narrower, slightly conical, top section, which was cast in a plastic paint container. The aluminium ex-telescope tube was partially buried in a 1 m (3 ft) deep hole that was excavated in the patio and then filled with concrete. Threaded rods were cast into the top, conical section to facilitate attachment of the mounting metalwork.

The new telescope tube is plastic. It was blackened inside by painting with matt black paint, and then emptying sawdust into the tube, so as to stick to the wet paint. When this had dried on, it was painted black again. A better secondary

Figure 9.28. The 245 mm f4.8 Newtonian telescope on its concrete and aluminium pier.

support system than the original was made, from various pieces of aluminium, including a 30 cm (1 ft) ruler that acts as a two-vane spider (Fig. 9.29). The original tube rings were thin aluminium and were replaced by custom-made stronger ones. The originals found a new life, clamped round the ex-tube, now pier, as an attachment point for shelving and electrics.

The current mounting is a Celestron CG5 GOTO model (that came essentially free with the C-11), but this is much too slight for the load now on it, and it will be replaced by something more suitable in due course. It was adapted to the pier-top using a cup-shaped aluminium disk, into which the mount base was bolted. This disk was screwed to another disk which was clamped down by means of flanges bolted to the threaded rods embedded in the concrete (Fig. 9.30).

For a while I used this telescope fixed, but without a proper covering, protected from the weather by a plastic car cover which was weighted down with stones. This was not too satisfactory, as it trapped dampness beneath it, and picked up

Figure 9.29. Curved Newtonian secondary support made out of an aluminium ruler, and new, centrally-sprung collimatable secondary holder.

a lot of dust and grit from the ground which it scattered onto the telescope when the cover was removed and replaced. Hence I needed to make a better covering that did not collide with the operation of the main observatory, but allowed both telescopes to be used simultaneously. The solution was a small run-off shed, approximately a 1.5 m (5 ft) cube, with double doors at the south-east end, that could run off the telescope and into the space between the brick columns supporting the run-off roof rails, fitting neatly underneath the opened roof of the run-off roof shed (Fig. 9.31).

Wishing to avoid having rails on the ground obstructing the observing space and getting tripped over by visitors in the dark, I conceived the idea of constructing a deck out of normal, grooved, decking timbers, the grooves running in the direction in which I wanted the observatory to run, and then relying on those to constrain the running of the shed, without the need for conventional rails. In the event this plan had to be slightly modified. The deck was constructed, and the shed, but in the interval between the construction of the run-off roof shed and the run off-shed, it had become impossible to obtain the 7.5 cm (3 in.) plastic wheels with rubber tyres I had used with such success before, and I had to

Figure 9.30. Adaptation at the top of the pier for the CG5 mount.

Figure 9.31. With both telescope sheds open, the run-off shed, behind the trellis, fits under the roof of the run-off roof shed

make use of 10 cm (4 in.) all-plastic ones. These proved to have poor transverse grip on the deck, and did not run quite on a predictable course.

I solved this problem as shown in Fig. 9.32, with some cycle components: small nylon cog wheels known as jockey wheels, which are mounted to the frame of the shed on steel brackets, on one side of it only, and run in a gap between adjacent deck boards. The edge of this gap, on the side on which it seemed to wear, was reinforced with a flat aluminium strip, of the type used for carpet edging at doorways. This strip does not create any night-time trip hazard, as a rail would do. The jockey wheels do not take any weight, but just keep the running of the shed on track. They return it, after use, to a position in which the doors bolt to the deck, to keep them closed, and also to prevent the shed from moving. One door also bolts into the underside of the roof, and slightly overlaps the other, to hold it closed.

The shed was constructed to fit in visually with the main observatory, and so used largely the same methods and materials, with jointed and screwed wall frames clad with featheredge timber. The frame was lighter (2 × 1 in. timber rather than 2 × 2), appropriate to having to move the whole shed rather than just the roof, but the featheredge wall construction was the same. The roof was done differently, being made of some surplus chipboard flooring boards that came out of the house loft. Three were used, and they keyed together using their manufactured tongue-and-groove design. They were covered with some green mineral felt left over from re-roofing the warm-room shed (really an old

Figure 9.32. Detail of the run-off shed, showing one of the running wheels, one of the jockey wheels, and the aluminium strip.

summer-house). The run-off shed, being three-sided, with no base, was reinforced at the bottom by diagonal planks (Fig. 9.33). The internal corners of the 2 × 1 in. frame were also reinforced with extra pieces of wood, all screwed through.

The wheels were installed in similar manner to the run-off roof wheels, mounted on M6 coach bolts, but these were mounted on extra blocks of wood attached to the frame, as the 1 in. nominal thickness of the framing timber was too slight to take them. These blocks were first screwed to the frame from below, but later, with time and damp, the construction "gave", and the gap between the base of the shed and the deck became too small, causing it to jam. The wheels had to be re-positioned, and at that stage the blocks were bolted to the frame, as seen in Fig. 9.32, for extra rigidity and to facilitate easier future adjustment.

The doors were made from tongue-and-groove timber about 1 cm (0.4 in.) thick, ledged top and bottom and diagonally braced with 2 cm (0.75 in.) thick timbers. They were hung directly on the outside of the side walls using four small butt hinges on each door, the hinges, on the wall side, being screwed to the frame through inverted pieces of featheredge, as with the other shed, and on the door side, screwed to the outside edge of the door timber. (The tongues and grooves of the door timbers were planed off at the ends that took the hinges. They were partially planed off to create the overlap between the doors.)

The roof was made with only a slight slope of 5 cm (2 in.) in 1.5 m (5 ft). The reason for this was that the slope of this roof is at right-angles to the slope of the run-off roof, to allow the run-off shed doors to be rectangular and full-shed height, and I wanted to maximise the height of the shed that could be fitted under the open run-off roof. This slight slope also simplified construction. In this case I did not bother to cut any timbers or joints to angles other than 90°;

Figure 9.33. Looking into the run-off shed, showing diagonal bracing.

the slope was accommodated in the margin of flexibility of the joints. This small slope has proved to be sufficient for shedding water. But one disadvantage of the whole concept is that the roof dumps its rainwater catch onto the wooden deck, and this cannot be easily mitigated – though little water seems to flow back under the shed walls.

In operation, the doors are first unbolted and unlocked, opened partially, and then the shed is pushed back by pushing on the top of the frame, and the doors are opened fully (the space available prevents the doors being completely opened without the shed being pushed back somewhat). The shed ends up abutting the south-east wall of the run-off roof shed, at which point the doors are secured in an open position using the large tower bolts at their lower edges, which engage with rings screwed to the edge of the decking. If this is not done, the doors can flap in the wind. To close up, these bolts are withdrawn, the shed is pulled back over the telescope, the doors are simultaneously closed, and the left-hand one is bolted to the deck using the same bolt, at its bottom inside edge. The right hand door closes last and bolts to the roof, and to the deck on the outside, and the doors are locked with a hasp and padlock. Were the shed located in a more exposed position, I would probably want to adopt further measures of securing it down, but it is very sheltered where it is. It is also heavy, and has shown no sign of being moved by the wind in the year it has been in position.

A small plastic table and an observing chair are stored on the bottom bracing planks of the run-off shed. The pier has two weatherproof mains sockets and corresponding switches attached to it, and these are supplied from the run-off roof shed via a cable running in a flexible steel conduit under the run-of roof shed floor and under the deck. One socket is used by a 13.8 V, 4 A power supply, which powers the mount, and the other is used by another of the small dehumidifiers. Before I installed this, the run-off shed, because of its small volume, suffered from a severe problem of trapped damp after observing nights, even more than the run-off roof shed, causing components to rust.

The third building constituting my observatory is the warm room shed (Fig. 9.34). This is actually an old summer house, which stands to the north-west of the patio only a couple of metres from the run-off roof shed. This shed has also been supplied with mains power points, so it can be heated by an electric convection heater. (I did try, before installing the electricity, heating it with a paraffin (kerosene) heater, but that proved rather smelly in the small space.)

The warm room is connected to USB hubs on both telescope piers via chained 5 m (16 ft) passive repeater cables, connected through a third, powered, hub. Hence a computer in the warm room can potentially control both mounts, plus imaging cameras and guide cameras attached to both telescopes (subject to sufficient bandwidth being available). Focus can also be controlled via USB through the Astro-Physics mount, and also on both telescopes via a DC cable laid to the warm room, leading to a simple reversible-current switching box, which duplicates the battery-powered hand units usually supplied with DC focus motors. I also have various other connections to the warm room which I have superseded, such as a serial extension cable. Other features of the warm room are charts on the walls, notebooks, a fire extinguisher, and an electric piano (for whiling away the time during long deep-sky exposures).

Figure 9.34. Inside the warm room.

There are limitations to my remote control system. Changing filters for tri-colour imaging with mono cameras still requires trips back to the telescopes, as I do not have a motorised filter changer. So do other common adjustments, such as changing the alignment of a guidescope with respect to the main telescope, to find a bright guide star. I tend not to use the warm room for imaging the bright planets, because, with the rapid rotation of these, time is at a premium in securing a series of images at different wavelengths, requiring rapid filter-changing. When one is operating from the warm room one can see the operation of the telescope in the run-off roof shed out of the window, but not the other telescope. Being so close is comforting, as there is less potential for something to go badly wrong before one has time to stop it than there would be in a system controlled from a more remote building, such as the dwelling-house.

One always learns from experience, and there are some features of my observatory that I would arrange differently if I were to start again. I would avoid the compromise on the shed depth owing to the length of the roof panels. I would use longer panels or double them up. In fact I think I would now start with a 3 m (10 ft) square run-off roof shed in view. With the larger telescopes than the original, more equipment, and occasional use of the observatory by more than one person at a time, its dimensions are too tight.

While I did plan for possible issues with the telescope pier by including a small area of removable floorboards around the original pier, this leeway was lost when I thickened-up the original brick pier, using all the space available between two floor joists. My brick pier was then hemmed in by immovable joists, which were screwed to floorboards by screws that became inaccessible after the walls were erected. I should have made the entire wooden floor removable from the inside, or at least a very large part of it, and I should have avoided trapping screws under the wall frames. When I finally decided to replace the bulky and elastic brick pier with a concrete one, I did put the floor back together in a slightly different manner, ensuring that most of it could be taken up again, if necessary, without having to remove the walls.

The lack of headroom in the observatory when it is closed is makes it difficult to work inside on rainy days, which is a thing that I have found I wanted to do quite frequently, and it also makes fetching equipment in and out, say to take to the other telescope, inconvenient, if the roof is on. Also the clearance on the telescopes is rather tight. The idea was to maximise the available low sky, but, in fact, there is rarely much point in observing below 20°, due to turbulence and absorption, save for special events, which may occasionally be visible low down, such as an eclipse or a comet. In the end, not much was gained by making it so low – I would make it a few inches higher if I did it again.

A minor problem with the run-off roof is the heating in the summer due to its translucency. After trying various remedies for this, I eventually painted the top surface with three coats of gloss white paint, but this is still only partially effective.

The basic idea of the run-off shed "interleaved" with the operation of the run-off roof shed works really well, as does the interchangeability of telescopes on mounts, an idea that I got from Dave Tyler. The observatory is very successful, in that it allows easy and comfortable operation, within a couple of minutes of leaving the house, of a choice of two large, well set-up and individually equipped, permanent telescopes, having very different focal lengths. The observatory also makes very efficient use of the available space, and no-one has complained about its appearance (Fig. 9.35).

Figure 9.35. The observatory in the winter, with both telescope sheds closed.

Dave Tyler's A Priori Fibreglass Dome

I have placed the words *a priori* in the title of this section, because this was the aspect of Dave's observatory that most impressed me on talking to him about it. It looks utterly professional, as if it was based on a detailed study of all the best dome-design precedents, amateur and professional. Yet Dave told me that when he designed it, he had not studied any other observatories, nor looked into the subject. He simply designed round the constructional techniques he knew he himself could use, and thought out the problem of the geometry and mechanism for himself, from scratch. The result is remarkably successful. Dave was able to do this because, as he says, engineering is in his blood. His father was an engineer, and Dave worked for most of his life in an engineering drawing office, latterly using CAD (computer-aided design) software. This experience gave him a semi-automatic feel for how engineering problems can be solved.

Dave's observatory stands at the end of a medium-sized garden, about 20 m (60 ft) from the house, in a large village in the Chiltern Hills, west of London. Dave is known as one of the finest planetary, lunar and solar imagers anywhere, and another of the finest, his good friend Damian Peach, lives just in the next village, in the valley below. This location is no rural idyll, however: the M40 motorway passes between these two villages, and only 200 m (600 ft) from Dave's observatory, and numerous other roads and railways follow the valley. The close proximity of two such successful imagers has caused much speculation as to "special atmospheric conditions" that might exist here, but this is unfounded. It is an ordinary English semi-urban environment with much the same conditions as the rest of Britain. What is extraordinary is the dedication of these two astronomers.

The observatory is 3.6 m (12 ft) in diameter externally. The dome consists of 10 fibreglass panels, each of which has two facets, so it is a 20-sided shape (Fig. 9.36). The drum-section is cylindrical, and consists of 8 curved fibreglass panels. The wall is only about 1.2 m (4 ft) high, and the door less than this, so, in common with most amateur domes, it is a bit of an exercise to get in. Once one is in, however, it is roomy. Dave did have a hardboard lining on the inside of the wall in the past, but recently he removed it to lay a new floor, and does not intend to put it back, as he likes the increased space.

The foundation for the dome was 10 cm (4 in.) of concrete, with a 1 m (3 ft) cube of concrete under the telescope pier. Joists were laid on the concrete, which in fact were old doorframes, and the new floor, consisting of waterproof plywood, further treated with Sadolin wood protector, is now laid on these.

The panels for the wall and dome were made using wood and hardboard forms. One form was used for all the dome panels. Each form was in two sections, to make it easy to separate it from the finished fibreglass. All the panels show a seam, where the two sections of the form separated. The forms were varnished and then waxed with slip wax. Fibreglass mats and resin were then built up on the forms to a thickness of 6 mm (0.25 in.). Silver pigment was put into the

Figure 9.36. Dave Tyler with his dome (and archery target behind).

fibreglass for the wall panels to make them opaque; the dome was left translucent white.

The observatory structure is unusual in that it essentially has no frame, either of wood or metal – one sign of Dave's independent thought. The wall panels were made with short side-faces, or flanges, at right angles to their surfaces, and these side-faces abut directly onto one another, and are bolted together, with metal plate reinforcements. At intervals, the wall panels have been strengthened by having vertical laths of wood incorporated into the fibreglass. The fibreglass has been folded over the laths, and they show as vertical ribs on the inside. The dome sections are likewise bolted to one another through their edge flanges, and have had waterproof tape (black in the photo) stuck over the joins on the outside.

The dome track consists of eight 16 gauge flat aluminium sheets bolted to the flanges on top of the wall sections. These were cut to the curves required using a band-saw (Dave has a well-equipped workshop, which has been crucial in fabricating components for his observatory and telescopes). Fixed castors bolted to the under-edges of the dome panels move on this track, and another set of castors, which retain the dome horizontally, are attached to the inside of the dome just above the lower edge, and bear on a sheet aluminium ring, again made from short sections, going round the top of the wall. It will be seen that no large metal components needed fabricating for this design: no forged circular rail, as is often considered essential for a dome, or similar. Dave has got round some of the most difficult issues of dome construction neatly and simply.

The dome has two shutters, which slide apart, moving on straight tracks at both ends, in the manner of the Mount Palomar observatory and others. They are not 100% waterproof: rain can be driven in by the wind from certain directions, hence the care taken to waterproof the new wooden floor. Creating completely waterproof dome shutters seems to be a very difficult, if not impossible, task. The shutter tracks are aluminium channel sections, held to the dome by solid alloy rods (Fig. 9.37). The rods have their ends angled to match the dome and track faces, and were tapped so as to allow securing to the dome and track. The shutters, made from fibreglass, move on PTFE rollers, which do not necessarily have to roll, as they are well-lubricated. The lower track lies below the bottom edge of the shutter, but the upper one passes through the shutter sides, near the apex of the dome, to securely retain the shutters.

Figure 9.37. The struts for the dome tracks at the top and bottom of the shutters.

There is nothing haphazard about Dave's constructions. The dome was carefully designed and drawn before the patterns were made, and Dave comments that after many years working in engineering drawing, it was only this dome design that caused him to understand the usefulness of radians (units of 57.29°) as angular measurement. Most of the dome components are a definite number of radians in their dimensions. He also comments that he never found the need to use radians again.

The observatory was constructed when Dave moved to this house, in 1977, and so has withstood 30 years of Chiltern hilltop weather, including the great storm of 1987, which damaged or destroyed many observatories in the south of England. The surface of some of the fibreglass is showing some slight crazing lines, in which algae are probably growing, and some of the galvanised fittings are slightly rusty, but the observatory shows few other signs of age.

Dave has had a number of telescopes in the observatory since it was built. He is perhaps best-known for the extraordinary images he has taken here of Mars, Jupiter and Saturn using Celestron C-11 and C-14 SCTs. The C-11 was, in fact, owned by Damian Peach previously, and has been called by others, jokingly, the WOMPOT – the WOrld's Most Powerful Optical Telescope, in consequence of the many wonderful images that have been taken with it. The idea that Damian and Dave's results have something to do with a specific telescope is, of course, another myth, like the special atmospheric conditions. Dave had much imaging experience behind him at this stage. In the early 70s, he was taking planetary images on film, using reflectors, from his garden in Slough. The observatory was constructed to house a large, skeleton-tube, double reflecting telescope of Dave's construction, consisting of two Newtonian systems – one having a 39 cm (15.5 in.) primary, the other a 21.6 cm (8.5 in.) primary made by George With in 1827. This latter mirror was one of the oldest surviving glass speculums anywhere. (Dave later exchanged it with a US astronomer for his C-14).

The large reflector housed in the observatory proved to be not too successful on planets. The mirror took a long time to cool down in an observatory in which the diurnal range in temperature can be large, and the skeleton-tube construction implied a long light-path through thermally unequilibrated observatory air. The smaller With reflector was better, and, with this, Dave started taking his first successful webcam images in 2004, after a period of visual observation using refractors. These instruments have now been displaced by the SCTs which he uses almost exclusively for planetary imaging. He sees these closed-tube instruments, with relatively fast-cooling optics, as being far better for high-resolution imaging, particularly in a domed observatory. To promote faster thermal equilibration of the observatory, he also now has fitted an extractor fan in the wall.

Recently, following a period during which none of the bright planets have been well-placed from here, Dave has become very interested in imaging the Sun, both in white light, and at the hydrogen alpha wavelength (655.3 nm). His instrument of choice for this is his huge 15 cm (6 in.) f15 refractor, and this is shown in the observatory in Fig. 9.38. However, Dave has engineered a system so that different large telescopes can be conveniently swapped on the German equatorial mounting. Each telescope has tube-rings attached to a base plate, fitted with two threaded studs at a separation of 26.7 cm (10.5 in.). These engage with holes in the saddle plate on the mounting. The tube-ring assembly is first attached to

Figure 9.38. The 6 in. f15 refractor in the dome.

the saddle plate without the telescope (after suitable counterweighting has been attached to the other side of the mount), the tube rings are held open by straps, and then the telescope is lowered in. For the C-14 and the big refractor, Dave says this is a ten-minute operation. Longitudinal balancing of the telescopes is also important, particularly in view of the heavy imaging equipment Dave uses. He has machined, in his workshop, an elegant arrangement consisting of a steel weight sliding on an aluminium bar, attached to the saddle plate (Fig. 9.39).

The 150 cm (6 in.) refractor was built by Dave using a doublet from Emerson Optical of London, a rack mount by Ronald Irving, of Teddington, Middlesex, and a 3 mm (0.12 in.) thick aluminium tube. This tube was too large to machine on his lathe, but he was able to work on it using a milling cutter attached to a power drill. Dave figured the objective himself, and determined the figure to be correct by star-testing it. (On other occasions he has had tested objectives using an artificial star created by light reflected off a ball-bearing placed as far away as possible.)

A domed observatory is the most suitable type for solar observing, since the dome blocks out most of the direct sunlight. Solar observing and imaging can be very awkward in the open or in a run-off. There is too much glare, and using a computer monitor is difficult. This can be partly remedied by some box arrangement to cover the observer's head, together with the monitor. Dave does in fact make use of a cardboard box as well, as his fibreglass dome is quite bright inside in daylight.

Hydrogen alpha imaging is made possible by a Daystar etalon-type filter which fits into the drawtube. This requires a power supply, as it contains a heater to

Figure 9.39. Adjustable counterweight and bar on the saddle plate.

maintain the filter at one precise temperature, at which it passes the H alpha line. It also needs to receive light from the objective in a cone at f30 or greater in order to optimise the selectivity of the filter for the H alpha waveband, and minimise the transmission of heat. Hence it is used on the f15 refractor with a 2× Barlow. The Daystar filter also has to be used with an energy rejection filter (ERF) mounted over the objective. Dave has discovered that these Schott glass ERFs transmit a significant amount of infra-red radiation, and that, for best results, an IR blocking filter is required in front of the Daystar as well. The final element in the imaging chain is a Lumenera LU75 USB 2 camera. This set-up produces superlative H alpha images (Fig. 9.40). Dave comments that a part of this performance must be due to the exceptionally great un-amplified focal length of the big refractor. There are few, if any, other observers using an f15 instrument for this type of imaging. The short, large-aperture refractors favoured by many observers today tend to suffer from spherochromaticism, which is significant spherical aberration at extreme ends of the spectrum, which can influence monochromatic as well as colour images. Of course, the use of such a large refractor is only practical in an observatory that is, by amateur standards, large.

Dave also carries out white-light imaging of solar features with the refractor. For this, he has constructed his own Herschel-type solar wedge. This sends most of the sunlight out of the back of a prism, away from the camera, and it is used in conjunction with various filters. It provides higher resolution than reflective-film methods of reducing the solar energy, such as the commonly-used Mylar sheets.

The telescope mounting currently in use in the observatory is a large German equatorial that Dave has built. It is simple, but extremely stable. The rectangular

Figure 9.40. A magnificent solar prominence imaged by Dave in hydrogen alpha light with the 6 in. refractor.

concrete pier is topped with a steel rectangular section (the height was increased with the change from Newtonian telescopes to refractors and SCTs), and bolted to this is a slab of synthetic resin-bonded paper, a hard, incompressible material not unlike thick Formica. The mounting base-plate is bolted to this. The main sections of the mounting were made from castings by Ronald Irving, which Dave machined. There is only a synchronous motor drive on the RA axis. Declination adjustment is by nudging.

The friction in the dec. movement is adjustable using screws which adjust the pressure on a slip-clutch arrangement concealed within the aluminium dec. setting circle. The friction in the RA movement is fixed. The RA axis turns on ball races, and a PTFE (Teflon) pad below the worm wheel ensures the friction below the wheel is less than that above, so the telescope is driven, but can easily be slewed manually.

Dave cut the RA gear himself by the method of "hobbing" on a lathe. This involves using a rotating tap, mounted on the lathe in a special attachment, to cut the edge of a rotating disk of metal. It is not easy to produce the correct number of teeth on a gear wheel by this method, but it is possible. Dave also engraved the RA and dec. circles himself using a dividing engine, and stamped the numbers on them. The RA circle is fitted with blocks front and back engraved with lines (Fig. 9.41). When these correspond, the telescope is on the meridian, and the local sidereal time is the RA of the meridian. This method of calibrating RA is used by Dave to find planets in daylight. Being entirely an observer of the Sun, Moon, and bright planets, he sees no need for GOTO facilities or digital setting circles.

The RA drive rate is only alterable in a limited way. To slow it, Dave switches it off, and to speed it up, he switches from mains current to the slightly higher frequency AC derived from an oscillator and step-up transformer. This crude system he finds sufficient for imaging the Sun, Moon and planets. If he were

Figure 9.41. Front and back of the RA assembly.

designing the mounting again, however, he thinks he would include a dec. drive. This cannot easily be incorporated into the mount now, but it would be useful to him for the purpose of producing image mosaics of the surface of the Sun and Moon. The dec. nudging method is a little crude for this. A very valuable feature of the mount, however, is the way the telescope can be turned most of the way round in RA without colliding with the pier, so minimising the need to reverse on the meridian (normalise). In this it differs from most modern GEMs. This is particularly valuable in the practice of high-resolution imaging, for preserving collimation.

The mount was accurately aligned, on first setting-up, using a theodolite and solar transit to establish the meridian in the observatory, shown by a mark on the wall, and no latitude for azimuthal variation was built in, other than the potential play on the bolts holding the mount down. Latitude adjustment is by means of a single M20 bolt, and dowels have been inserted into the vertical part of the mount base to hold it for more accurate adjustment of the latitude setting (Fig. 9.42).

Dave likes making his own equipment, and is one of a rare breed now, though he is not sentimental about it: he recognises that times have changed, and that many things can be far better and more cheaply mass-manufactured today than ever they could be before, and that this naturally has limited what it is sensible to try to do oneself. He saw the first BBC *Sky at Night* broadcast in 1957, and, soon after, read a book on telescope making. He subsequently ground various mirrors, re-figured lenses, and made optical test equipment including a Foucault tester. This work culminated in the construction of the 39 cm (15.5 in.) reflector and a 21 cm (8.25 in.) folded refractor, but he realised that the latter was an outdated, unnecessary idea when he compared it to the similar-aperture catadioptric telescopes coming onto the market in the late 1970s, which could provide similar performance in a far smaller and lighter package.

In 2004 he started to use the Philips ToUcam colour webcam. Initial experiments, using it afocally with refractors, were not a great success because of the lack of colour correction of the doublet objectives across the range of sensitivity of the CCD. This prompted a return to the With reflector. At the same time, Dave learned, with Damian's help, how to use Cor Berrevoets' *Registax* image stacking software, and Adobe *Photoshop,* to process images. This set-up was succeeded by the Atik Instruments ATK 1HS monochrome webcam, used with a C-11 and

Figure 9.42. Base of the mount.

filters, to build up a more detailed colour image than could be created with the "one shot" ToUcam, by imaging in each colour in quick succession, stacking and sharpening the monochrome images in *Registax*, and then combining them and adding a luminance layer in *Photoshop*.

The Atik USB 1 camera was succeeded by the Lumenera LU 075 USB 2 camera, which offers faster frame rates without data compression. It will run at up to 100 frames per second (fps) on reduced-size fields, or 30 fps on the whole 640 by 480 pixel field. The primary advantage of the faster frame rates is that they allow the collection of more data in the short period available before a planet has rotated too much. The faster frame rates, however, can only be used on objects that are sufficiently bright – Saturn, for example, has a low surface brightness, and is only imaged at 15 fps. Dave usually uses the Lumenera camera with the C-14 for planetary imaging. The success of the C-14 here is probably not wholly down to the increased resolving-power of a larger aperture. It is rather that the large light-gathering power of this telescope allows a very large image scale to be used while keeping the brightness, and signal-to-noise ratio, high. The processing software is able to much better perform its tasks of aligning images, and sorting them in quality order, if the image is kept bright and the signal-to-noise ratio high. Additionally, more data can be collected from a brighter image, because the camera can be run faster. Thus, larger apertures have a triple or quadruple benefit in webcam imaging.

In 2005 and again in Summer 2006, Dave and Damian both took their C-14s, Celestron CGE mounts, Lumenera cameras, computers and other equipment to Barbados to capture the southerly-declination oppositions of Jupiter in those years, which could not be well seen from the UK because of the low altitude. They also took many images of the other planets while they were there. The seeing conditions on Barbados proved to be generally superb, and an amazing body of work was built-up, particularly covering the development of atmospheric features on Jupiter. One result of this work (co-ordinated with observations by others in different parts of the globe by John Rogers of the BAA) was a new determination of the rotation period of the Great Red Spot, which had previously only been accurately measured from images sent back by the *Voyager* probes.

Dave's excellently-equipped workshop is key to his endeavours. He built this himself, utilizing some metal doors and windows that were being thrown out by a neighbour. It is an L-shaped shed to the west and north of his observatory, containing a lathe, drilling machines, a milling machine, mechanical saws and metal cutters, grinders, and much else, including an impressive selection of other out-of use telescopes and telescope parts. Dave can make pretty much anything in metal, or other materials, provided it is not too big. Among the smaller items of his making that I saw is an adaptor, ingeniously baffled by having a sharp screw-thread cut into it internally, which makes the un-painted, un-anodized metal virtually non-reflective from the point of view of the optical path.

Dave has put his workshop skills to non-astronomical uses, such as the making of cross-bows for archery. This has been his other long-standing enthusiasm, and he has been British field archery champion with the crossbow and recurve bow. One can see the connection between the accuracy required in this sport and his dedication to precision in manufacturing astronomical equipment and acquiring world-beating images with it. His attitude is that once he tries his hand at something, he has to do it as well as it can possibly be done, and it is this that drives him to push back the boundaries of amateur high-resolution astronomical imaging.

Norm Lewis's Observatory: An Experience with a Commercial Dome

Norm's beautiful observing site in Maryland, USA, is shown in Fig. 9.43. This rather open location has the disadvantage of being windy, and, in the winter, temperatures can fall very low – to below -18°C, zero degrees Fahrenheit. For these reasons, Norm opted for a domed observatory, and purchased the 10 ft ProDome from Technical Innovations, also based in Maryland. The ProDome is constructed from four fibreglass quadrant sections, and has an "up and over" shutter.

The ProDome design offers a solution to the common problem with amateur-sized domes that the door is too low to get through comfortably. The dome does not have a continuous base ring. The base ring is interrupted at the wide shutter, which is 1 m (3 ft) wide, and the dome support ring, on top of the dome wall, is continued in a short section forming the top of the low door. Thus, when the door and shutter are opened and coincident, the dome can be entered though

Figure 9.43. The situation of Norm's observatory.

the combination of the two, without having to stoop. When the door is shut and the shutter is open, the dome can be rotated all the way round on its track.

A disadvantage of the wide shutter is that it does allow condensation in the dome on still nights. Norm has countered this with the installation of a dehumidifier (Fig. 9.44). This has a temperature sensor, and is set to run only if the temperature is above 2°C (35° Fahrenheit). The prevailing wind is from the west, and the exposure is also greatest in this direction, with few trees. There are some trees to the north, but south and east are almost tree-free. Consequently, Norman prefers to observe to the south and south-east, so that the dome blocks out most of the wind.

The main telescope is currently a Takahashi Mewlon 25 cm (10 in.) Dall-Kirkham Cassegrain, used largely for visual observing. This is shown in Fig. 9.45 on an Astro-Physics 900GTO mount, mounted by way of tube-rings and a Robin Casady dovetail system. A Takahashi 10 cm (4 in.) refractor is also mounted to the tube rings, by way of a Losmandy dovetail and mounting ring system, opposite to the declination axis. This secondary instrument has since been replaced with a larger one, and the mounting has been upgraded to the larger Astro-Physics 1200 model. Two 8 kg (18 lb.) and one 4 kg (9 lb.) counterweights are used to

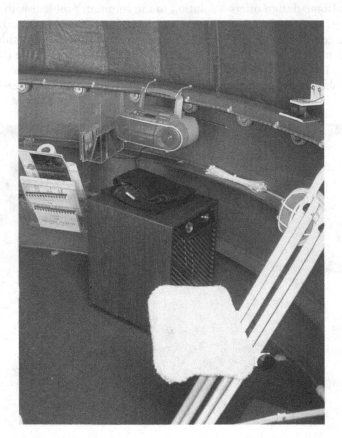

Figure 9.44. Some of the internal furnishings, including dehumidifier and observing chair.

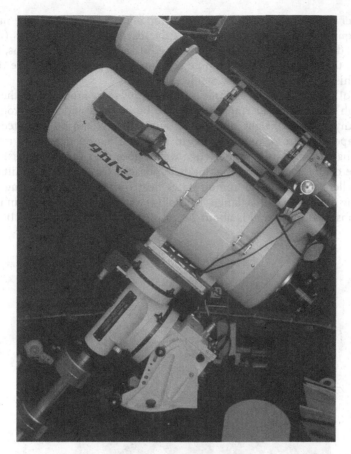

Figure 9.45. The Takahashi Mewlon reflector and Takahashi refractor on an Astro-Physics mount.

balance these instruments. This combination of weights allows the flexibility to add heavy accessories like a binoviewer and compensate with a small adjustment, or to change a whole OTA, and make a large one.

For about a year Norm used a Meade LX200 40 cm (16 in.) SCT in this observatory, on the AP900 mount. This was a successful combination, but he found the telescope a bit too large for the observatory, and ultimately went over to the smaller Dall-Kirkham, which he finds to be a superb instrument for visual planetary observation, though he misses the light-grasp of the big SCT for seeing faint objects. The smaller telescope does leave enough room for the observer in the dome, however, which is rather important for visual observing. (Note how much larger a 10 ft dome is compared to the 7 ft domes offered by some companies: it has twice the floor area.)

The mounting is bolted to a 20 cm (8 in.) diameter steel pier that extends through the wooden floor, and this is bolted to a 3,500 kg (3.5 ton) concrete block 1.2 m (4 ft) square on its top surface, that goes 1.8 m (6 ft) into the ground. The deck was engineered slightly higher than expected when the pier was ordered, so the pier ended up being 7.5 cm (3 in.) too short. Norm raised the mounting to the required height using four short galvanised steel pipe sections (nipples). Bolts run down from the mount base plate, through the pipe sections, to holes tapped in the pier top. Since the pipe sections were pre-cut, they are the same length. This arrangement is very stable (Fig. 9.46).

The ProDome is made by Technical Innovations to be mounted on their 30 cm (1 ft) high fibreglass wall rings, which can be stacked and bolted together to achieve any required height. The dome, alternatively, can be mounted on any other shape and size of building. Norm mounted his on three wall rings. The top ring has a flange at its upper edge which goes over the dome base ring on the inside of the observatory, making it impossible for the dome to lift off. The rubber rollers on which the dome rotates bear on the dome base ring through gaps in the underside of this flange. The rollers and flange can be seen in Fig. 9.44. Norm's observatory overall is about 2.4 m (8 ft) high from the deck. It is raised

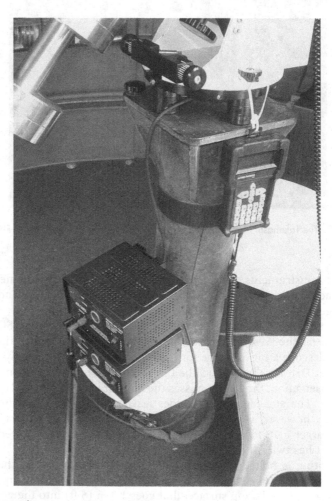

Figure 9.46. Telescope pier, mount controller and power supplies.

above the ground on the sloping site by a deck which varies between 20 cm and 60 cm (8 and 24 in.) in height above the ground. To access the square platform from the low-ground side, Norm constructed a short flight of steps and an elegant balustrade (Fig. 9.47).

A significant feature of this observatory is the lightning-protection system that Norm has built in. On such an open site as this, lightning is a very real risk. The 6 m (20 ft) lightning conductor can be seen in Fig. 9.48. However, specific to fibreglass domes in open locations like this, there is an additional risk that one might not guess – that of electrical discharge from the dome to the sky (reverse lightning). This phenomenon was recognised after several fibreglass observatories in the USA were struck. After some research, it was discovered that the problem was the same as one that had previously been encountered with fibreglass gliders. As the gliders moved through the air, they could build up a static charge resulting in eventual discharge to the clouds. This problem was solved by glider manufacturers by building in a fine copper mesh between the glass layers, cross-connected during assembly, and routed to static dissipators on the wing tips and vertical stabiliser. Fibreglass observatory domes experience the same static build-up when wind blows continually over them under certain atmospheric conditions. The dome manufacturers, however, have been reluctant to admit there is a problem here, and, unlike the glider makers, have not changed their methods of manufacture.

Figure 9.47. Platform and steps.

Figure 9.48. Lightning conductor.

It is therefore necessary to ground a fibreglass observatory in a open, windy situation like this effectively. Norm has painted the inside of the dome and the wall with a dark blue, electrically-conducting latex paint. The conductivity of this paint is quite small, but it is enough to dissipate static before it becomes a major problem. The dome wall is connected to ground, across the wooden platform, by a thick braided cable, cross-connecting to the lightning conductor with a heavy connecting clamp (Fig. 9.49). The problem remains getting a connection between the dome and dome wall, which are separated by insulating rubber rollers. Norm has solved this problem with the white brackets attached to the base of the dome inside, one of which is seen in Fig. 9.44. When the dome is closed, the copper loops below the brackets contact with aluminium plates on the top of the dome wall flange. Thus, when it is closed, the whole dome is grounded. The dome is never in open configuration when there is bad weather.

Norm is largely a visual observer, producing fine drawings of the planets, such as those of Mars shown in Fig. 9.50. The other "personal" features of the dome shown in the pictures mainly add to the comfort of the visual observer. The seat

Figure 9.49. Earthing cable.

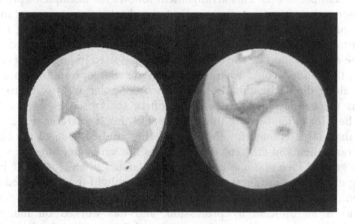

Figure 9.50. A couple of Norm's drawings of Mars.

is an old one, to which Norm stapled a piece of lamb's wool as a cover. This prevents dew forming on the seat, and him from slipping off. A plastic screw and nut holder attached to the inside of the dome is used to keep filters handy. The white dome lights on the wall are in fact red, with a dimmer to produce a very low level of lighting if required. There is also a white rope light all round the inside of the wall-top flange, to provide light for working inside the dome (natural light has been excluded by the anti-static paint). There are two separate 12 V power supplies on a shelf at the base of the pier, one for the mounting, and one for dew heaters. This ensures that the mounting does not experience fluctuations in voltage when dew heaters are turned on and off. An electrical conduit runs round the inside of the dome wall to supply further power points. One other small feature Norm has added, to combat the disorientation often experienced in a dome, is that he has painted the letters N, S, E and W on the wall inside.

The observatory has a name, as all observatories should, and a dedication, both of which are engraved on a copper plaque on the balustrade outside. The name is Norlin Observatory, a combination of the names of Norm and his wife Linda, and a tribute to her understanding of his regularly coming in at 3.30 in the morning. The dedication is "To the Fires of the Night", a quotation from the writings of Percival Lowell, who got it in turn from Christiaan Huygens. The true astronomer, Norm considers, is, indeed, one who is "captivated by the Fires of the Night".

Es Reid's Solar Observatory

In solar work, there is a less-frequently used alternative to the conventional large telescope on an equatorial mounting in a dome, as exemplified by Dave Tyler's setup. Es Reid's setup, based on a heliostat and fixed spectrohelioscope, is a good example of this. This arrangement follows the example of professional astronomers, who often use such complicated and extensive equipment to analyse the light of celestial objects that it cannot be carried on a moving telescope. The solution, then, is to arrange for the instrumentation to be fixed, and the light to be directed onto it. The arrangement often used in large observatories is known as the coudé layout,[5] where mirrors in the telescope and mounting direct the light along the declination axis and down the polar axis, to the fixed point known as the coudé focus, where, most often, spectroscopes are used. There tends to be a lot of light lost in this process, but this is tolerable if the telescope is more than large enough for the object being studied. Due to the light loss, coudé arrangements in amateur observatories are very rare, but one variant of the same general idea that it is possible for amateurs to employ is the heliostat, a moving mirror that produces a static beam of sunlight that can be fed to other equipment, and used for any desired purpose.

Es Reid is an optician by profession, and has worked on many of the most sophisticated optical systems employed by amateur and professional astronomers, as well as by industrial companies, the military, and other security forces. He designs and builds telescopes and makes all sorts of optics, work that is done by a combination of computer modelling, handicraft, and simple mechanical aids, and, which is, despite the predictive powers of modern optical design software, still something of an art. His problem with astronomy is that, like so many of us, he doesn't like getting cold. This is a major problem on the bleak East Anglian plain of England, where the wind usually cuts like a knife.

---

[5] Do not ask who Coudé was – the word is just the French for "bent".

His solution is to concentrate on observing the Sun from inside a completely enclosed shed. The heliostat directs sunlight into the shed through a hole cut in the wall, about 15 cm (6 in.) across. The hole is blanked off, when not in use, by a shutter. The heliostat (Fig. 9.51) consists of an optically-flat mirror 20 cm (8 in.) in diameter, supported on an altazimuth mount that used to belong to a Meade LX200 telescope. This was modified into its present form by Brian Brooks of Astroparts, Milton Keynes. Es did use an equatorial mount in the past, but the current arrangement is more stable. The azimuth gear of this mount has been replaced by a better one (made by Beacon Hill Telescopes), and both axes have been fitted with powerful new stepper motors supplied by AWR Technology. The drive electronics box for them can be seen on the plinth below the mount. The whole is supported on a 15 cm (6 in.) aluminium column embedded in gravel and concrete. The column is abut 1.3 m (4 ft) high. To install it, a 60 cm (2 ft) deep hole was dug, and a wooden fence post was concreted in. The aluminium column was placed over the fence post, and more concrete and gravel was added round it. The space remaining inside the column was filled with sand to damp vibrations. When not in use, the heliostat is protected only with a plastic bag tied over the top of the column.

Figure 9.51. Heliostat.

The tracking of the heliostat is controlled by a handset and a bespoke software solution supplied by AWR. The motor controller is connected to the handset, in the shed, by a long Category 4e cable. The power supply for the motors is also in the shed. The shed is a perfectly ordinary garden shed 2.4 × 3 m (8 × 10 ft) with the windows covered to make it dark inside. All the equipment in the shed, including the telescope, spectrohelioscope, camera and computer, is arranged on a laboratory-bench type table. This has steel legs, which are placed on concrete blocks (Fig. 9.52). The blocks go through holes cut in the floor of the shed, and are concreted to the concrete pad on which the shed stands, which is about 15 cm (6 in.) thick. The shed is thus mechanically disconnected from the table, and there is no detectable movement of the optics.

The equipment on the optical bench in the shed much more closely resembles an optical laboratory setup than conventional astronomers' equipment, and is easier to understand diagrammatically than from photographs, or from actually seeing it. The telescope is based on an 11 cm (4.5 in.) f15 oil-spaced doublet lens. This was originally made by the famous English optician Horace Dall. The flint glass element of it is adjustable, by screws, with respect to the crown glass element, to compensate for atmospheric dispersion (the differential refraction of colours, most apparent when an object is low in the sky). Both elements were re-figured by Es to be optimal for the wavelength of hydrogen alpha light. The rest of the telescope is a plastic drain-pipe, which has no function other than to shade the objective. The objective is moved back and forth to focus the telescope; the cell is mounted on a block moved by a threaded-rod mechanism, turned by a cardboard tube that just acts as a long handle. A short way along from the objective, there is a side-tube connected to the main telescope tube,

Figure 9.52. Base of the optical bench.

containing a 45° flat mirror and a negative lens. This acts, with the objective, as a Galilean telescope. It gives a wide-field view of the Heliostat mirror, which has sometimes been used for locating Venus in the daytime. This, however, is incidental (Fig. 9.53).

There are two main ways in which the telescope can be used on the Sun: for projection onto the far wall of the shed, or as a feed to the spectrohelioscope. Projection along the length of the shed produces a large solar image – 75 cm (30 in.) across. A piece of paper is hung from the roof in such a way that it can be swung like a pendulum. Swinging it helps to remove the grain of the paper from visual perception, making the solar details more obvious. A circle is drawn on the paper. The atmospheric refractive flattening of the Sun's disk in the winter is very obvious against this circle. However, dispersion can be almost removed by adjustment of the objective, so keeping the white-light image sharp.

The main point of the observatory, however, is the spectrohelioscope, which occupies an area of the bench alongside the telescope light path. The word spectrohelioscpe is confusing, as it is not a type of spectroscope, though it has much in common with one. It is a device for imaging the sun in any desired single wavelength. It is thus an alternative to the etalon-type solar filters such as the Daystar, used by Dave Tyler, or the popular Coronado solar telescopes. These are fixed to work in only one wavelength, normally hydrogen alpha or calcium K. To examine another wavelength, you have to buy a different filter or another complete solar telescope. This is very expensive. Es Reid's shoestring spectrohelioscope, however, cost something like £400 to make, and can be used to isolate any solar wavelength and image in it.

Figure 9.53. This view of the equipment at the objective end of the bench shows the shed -hole cover, the plastic telescope tube, the right-angled Galilean section at the back, the cardboard tube focusing handle, and the tuneable diffraction unit in front of the telescope.

The design was due to Brian Manning, and those interested can follow up the original reference.[6] It was further refined by Norman Groom, who previously had this spectrohelioscope set up in his roof, and by Es. Fig. 9.54 shows in essence how it works. A flat mirror at 45° to the light path of the telescope directs light towards a prism, with two aluminised faces at 90° to one another. The first of these faces deflects the light along a path parallel to the telescope, and of almost the same length, to a diffraction unit. This unit consists of a collimating lens, figured by Es, and a reflective diffraction grating 60 mm (2.4 in.) square, made by Brian Manning. The lens is necessary because the grating must receive and transmit parallel light for the system to work. The diffraction grating, set at not quite a right-angle to the main axis of the instrument, reflects a spectrum of the solar radiation back to the second face of the prism, where it is again deflected through 90°, into either an eyepiece or a camera.

In fact, a spectrum is not observed, because of the pair of connected slits, one on either side of the prism (Fig. 9.55). These isolate one colour, or spectral line, and produce a vertical line of light at the detector. If the first slit were not present, there would not be a sharp spectrum, because of the angular size of the solar disk, and if the second were not present, the whole spectrum would be observed at once. But because they are vertical slits at the focus of a telescope, and because the lines of the diffraction grating are also orientated vertically, it should be realised that what is actually observed is a narrow, vertical section of an image of the solar disk in one colour. Now, the slits are connected to each other and to a moving coil system, similar to that of a loudspeaker, which is controlled with electronics to cause the slits to oscillate laterally. When the slits are moved laterally though a small distance, the colour that is isolated remains constant, because angles of incidence, reflection and diffraction remain

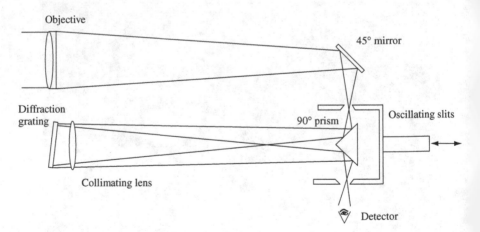

Figure 9.54. Principle of the Manning spectrohelioscope.

[6]Journal of the British Astronomical Association 1982, Vol. 92, No. 3, p 112.

Figure 9.55. This view of the equipment at the camera end of the bench shows, from the left, the 45° mirror, the oscillating slits to either side of the 90° prism, and the pear-shaped ToUcam. Just below that can be seen the control unit for the heliostat mount.

almost constant, but the vertical slice of the solar image they transmit to the detector moves across the solar disk, as the slits move across the focus of the objective.

If the slits are made to oscillate fast enough, and the eye is placed at an eyepiece at the final focus, persistence of vision will make the travelling slice of the monochromatic solar image appear as a complete disk. Note that no other filtration of the solar radiation is required in this system, because the slits only transmit a tiny proportion of the light from the objective. Es has been adapting this system to webcam imaging. A relay lens is used to amplify the image for a Philips ToUcam, used at 320 × 240 resolution. The complication with this initially was found to be that the webcam scanning rate could interfere with the scanning rate of the slits to produce image artefacts. This problem has now been overcome, by finding the optimum slit scanning rate, and some results are shown in Fig. 9.56 for both H alpha and Ca K light. The ToUcam is a colour camera, which produces a red image for H alpha and a purple one for Ca K. An improvement could be obtained by using a monochrome camera, which would be more sensitive, lacking the Bayer colour matrix on the chip, not needed for monochromatic imaging.

The slits are actually razor-blades, and their distance apart is adjustable to suit different viewing conditions. The slits also have to be kept very clean. The frequency of oscillation is about 20 Hz, but the system is effectively a pendulum, whose frequency can be modified by changing its mass, which Es does by adding bits of Blu-Tack and coins. Blu-Tack was also found to be useful for preventing a resonant response of the telescope tube to this oscillation. (All round, Es finds

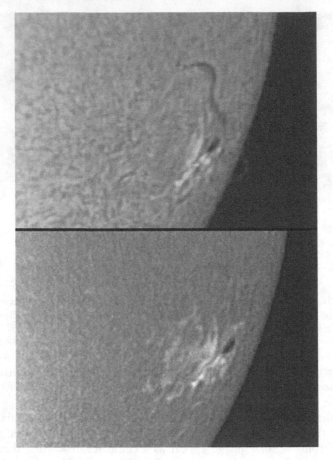

Figure 9.56. Corresponding hydrogen alpha and calcium K images taken by Es with his spectrohelioscope, 18 minutes apart.

it an invaluable substance for the optical engineer.) The frequency of light to be examined is changed by adjusting the angle of the diffraction unit. This shifts the whole spectrum slightly with respect to the slits, and thus changes the colour of the virtual image. The adjustment is precise, and is done with the slow-motion rods shown in Fig 9.53. The collimating lens is a singlet, and needs re-focusing for each colour.

Attached to his house, Es has a garage, which, like many astronomers' garages, is not used for storing a car. It has been converted into an optical test-tunnel 6 m (20 ft) long, lined with rock-wool to prevent thermal variations along its length. This he uses for testing mirrors, lenses and optical windows by double-pass against a 62 cm (25 in.) mirror, which is flat on one side and concave on the other. To me, one of the pleasures of visiting amateur astronomers' homes is coming across such extraordinary setups completely hidden behind bland, normal house-fronts.

Richard Miles' Compact Remote-controlled Photometric Facility

It has been generally insisted so far in this book that observatories need to be big enough to accommodate, comfortably, all the equipment that is to be used, plus at least one observer. However, this need not be the case, if the observatory structure is only to be a shelter for the equipment when it is not in use, never needing to accommodate an observer. Run-off sheds are one example of this type of shelter, but there are others. Two of these will form our final two case-studies, exemplifying respectively the high-tech and the low-tech ends of the observing spectrum.

Richard Miles is a serious observer of comets, asteroids and variable stars, particularly interested in making photometric observations. He moved a few years ago to a thatched cottage in a village in north Dorset, in the rural south of England, in search of darker skies. In his previous location he had used a run-off roof observatory, but his wife did not want such a large construction here to detract from the beauty of the cottage garden (as she considered). The solution arrived at was a minimum-sized enclosure that only allows space for the telescopes and computers, with operations always being remotely controlled from the house (Fig. 9.57).

The telescope cover was made by Richard from 13 mm thick plywood, glued and screwed. It is in two sections. The lower section is 90 cm wide by 81 cm deep by 73 cm high (36 × 32 × 29 in.) . The base is made from two layers of 18 mm (0.7 in.) plywood bonded together, and coated with epoxy resin-based flooring compound on both sides to make a watertight seal. The walls have a 3 cm (1.2 in.) lip all round the base to allow ensure water runs off easily. The plywood base is mounted on a steel baseplate, which was made out of welded steel strips by the village blacksmith. The idea was to make it impossible for water to get into the enclosure from below, and also to minimise the incursion of small animals, especially woodlice, which are so prone to crawl through any available gap in an observatory. Hence this baseplate was mounted on four 20 mm (0.8 in.) bolts set in concrete (Fig. 9.58). The bolts are daubed with grease to dissuade climbing insects. They raise the baseplate 12 cm (4.8 in.) above the concrete. A pipe was set into the concrete, which acts as a conduit for two mains electricity cables and two ethernet (category 5e) cables, which run to the cottage.

The upper section of the enclosure is 102 cm wide by 94 cm deep by 78 cm high (41 × 38 × 31 in.). This upper section is attached to the lower via flanges on either side which are attached to cantilevered arms on the sides of the lower section by means of 15 mm (0.6 in.) steel bolts which, rotating in steel pipes, with large circular brass washers, act as hinges (Fig. 9.59). The flanges in the upper section are extended into arms, beyond the hinges, to which counterweight boxes are attached. The boxes extend from the hinges about 50 cm (20 in.), and are filled with cut sheet lead, creating a total counterweight of about 40 kg (88 lb.).

Figure 9.57. Richard Miles' compact telescope cover.

The total height of the enclosure is 153 cm (60 in.). Concrete slabs are laid round three sides of the concrete base. The paving on the fourth side is lower by about 12 cm (5 in.) and has a rubber mat positioned so that when the upper section of the enclosure rotates on its hinges, opening up the observatory, it comes to rest on the mat in a position that places it below the horizon as seen by the telescopes.

The upper section is lined with 25 mm (1 in.) thick expanded polyurethane sheets normally used for insulating cavity walls. These have aluminium foil on either side. Their purpose here is to minimise the condensation that collects on the internal surfaces of the enclosure when it is open at night. Several trays of granulated silica gel are left inside, after closing up, to absorb moisture. About every two weeks these trays are regenerated by leaving them in the "Aga" stove in the cottage for several hours. The quantity of gel used has the capacity to absorb 600 ml (one pint) of water.

Figure 9.58. Steel baseplate supported on bolts.

Figure 9.59. View of the counterweight side of the enclosure, the base and the rubber mat on which the upper section rests when the observatory is in use.

The observatory is dedicated principally to photometry of asteroids and variable stars. The main telescope is a Celestron C-11 28 cm SCT, fitted with a 10 × 40 mm finder. Either side of it, mounted directly on to a wide saddle plate fitted to a Vixen Atlux mount, are two Takahashi FS-60C 60 mm (2.4 in.) aperture, f5.9 fluorite refractors. These two telescopes enable accurate photometry, as they are equipped with filters, one having a V-band filter, the other an I-band filter. These two filters are used to standardise brightness measurements – V stands for visual, and is centred on the green part of the spectrum, while the I is a near-infrared filter. The C-11 is used unfiltered, so as to maximise signal-to-noise ratio when imaging faint targets such as small, fast-moving near-Earth asteroids.

All three telescopes are equipped with Starlight Xpress SXV-H9 cameras, each connected to a laptop computer in the enclosure via a USB 2 interface. The field of view of the cameras is 11.4 × 8.4 in. for the SCT and 87 × 54 in. for the refractors. The C-11 and the V-filter refractor are also equipped with Robo-Focus focus motors controlled via the serial ports of the PCs. These PCs can themselves be controlled from another computer indoors, via the ethernet link. Each telescope has heaters fitted, not only around the front element to prevent dewing, but also around the adapter between each CCD camera and telescope, so as to prevent condensation or frost forming internally, close to the camera. This can otherwise be a problem with Peltier-cooled cameras. The cameras are turned on about 15 minutes after the heaters are switched on to prevent frost forming on any optical surfaces. The main telescope also has a flexible dewshield which is slid back over the tube before the observatory is closed up.

The Vixen Atlux mounting is on a pedestal base, and this stands on the base of the lower section of the enclosure. The final adjustment of the polar alignment, to within a few arcminutes of the celestial pole, was achieved by adjusting the four nuts supporting the entire observatory. The counterweight shaft can be retracted by unlocking a clamp on the axis and sliding it inwards: this allows the enclosure to be smaller than it otherwise could be. The mount is normally controlled remotely over the network, from a desktop computer in the study of the cottage running *Windows 98SE* and the planetarium program *Guide 8*.

For unfiltered photometry of asteroids and other objects such as gamma-ray bursters, all three cameras are run simultaneously. The wide fields of the small telescopes, about 1.5 square degrees, usually contain stars of accurately-known magnitudes, and these can be used to calibrate comparison stars within the field of view of the main telescope. Since stars are imaged using two filters, the difference between the two magnitudes obtained (M_v-M_I) provides a measure of the colour of the comparison stars, and allows unsuitable ones, such as very red stars, to be weeded out. The system can be used to carry out absolute photometry, by measuring atmospheric extinction to a high precision, and it can be used to determine the magnitudes of the stars in a comparison sequence to a lower limit of M_v=13.

Speed-of-use is one of the chief features of Richard's observatory. It can be opened up in a few seconds, similarly closed again if rain threatens. Once the heaters and cameras have stabilized, it takes about five minutes to power-up the computers in the observatory and calibrate the mount. After opening up, the trays of silica gel are brought in and covered up. At the end of the session, Richard normally comes out again to cap all three telescopes in order to take "dark frames" with all three cameras (these are used to help even out the variations in response across the camera detectors in the image analysis). Images taken with the cameras are stored in the outdoor computers during observing, but can be downloaded indoors at any time, to be stored on an external hard drive in the study. At the end of an observing run, the outdoor computers are turned off remotely, two of them being placed in "hibernate" mode in order to make starting-up quicker on the next occasion. The silica gel is returned, the cameras and Robo-Focus devices are switched off, and, from indoors, the electricity supply to the whole observatory is shut down.

Other uses of Richard's set-up have been to image the bright Comet McNaught of early 2007, Venus, and bright stars in daylight, using one of the refractors, with the objective capped with a very dark neutral filter to cut down transmitted light by 99.999%, or 12.5 magnitudes.

Apart from the computer technology, the main factor that makes possible a very compact observatory enclosure such as Richard's is the use of modern compact telescopes. A hinged box of this type would not be very practical with a traditional long-focus reflector or refractor. It is also a solution suited to particular types of observational work, but not others. For example, my experience has convinced me that remote-control high-resolution imaging of solar system objects is not a practical proposition, because the amount of sensitive adjustment of the instruments required every time makes it, ultimately, an unavoidably "hands-on" operation, which does entail being in the observatory, getting cold. This will continue to be the case until rather more sophisticated telescopes than are currently usual for amateur use become the norm. Richard admits that he rarely collimates his telescopes, as it usually makes little difference for photometry. The computerised mini-observatory is a clear sub-genre of the species now, that is growing in popularity.

Mike Morrison-Smith's Dobsonian Storage Box

Finally, as an antidote to all the automation and expensive electronics, Mike Morrison-Smith's telescope shed is shown in Fig. 9.60. This is definitely the cheapest observatory of those featured in this book, and also has the smallest footprint. The telescope is a 30 cm (12 in.) Dobsonian, that is borrowed from his local society. The shed, also known to Mike's wife as the "sentry box" (and he admits he does sometimes stand in it), was made from three internal doors, for the front and sides. The back was made from two 4 × 2 in. planks bolted to a concrete post behind. The sides and back are covered in roofing felt. The roof consists of the baseboard of an old shed, covered with some damp-proof material Mike found in a skip. The floor consists of some concrete slabs, that Mike says, "were lying around, as slabs do". The door was painted with grey paint given by a friend. "Come to think of it," says Mike, "there was a charge, as I paid for the tacks to nail the felt and damp-proof material on".

Figure 9.60. Mike Morrison-Smith's Dobsonian "sentry box".

The telescope is used by dragging it onto the concrete slabs at the front of the box using a sack barrow (a two-wheeled trolley with a short horizontal shelf at the front). This arrangement avoids the problem of having to get the telescope over any threshold or lip: simple, but brilliant. Mike is a dab hand at sometimes zany invention in the best tradition of amateur astronomical contriving. Perhaps his finest inspiration was his "garden curtains", to shield the street lights to the rear of his garden. He raised a rope across the garden at a level of about 3 m (10 ft), tied to a tree on one side, passing over a length of angle iron he inserted in the ground on the other side, and terminating at a post about 1 m (3 ft) from the ground. The curtains were made from black PVC sheet, and were suspended from the rope with cable ties going through 2.5 cm (1 in.) dowels. In the day he would draw the curtains to one side, and at night he would pull them across to observe. I do not know what fate eventually befell these curtains, but I like the story. Amateur astronomers are supposed to be eccentric, and some of them live up to the reputation.

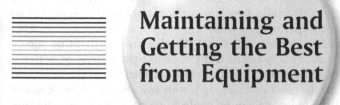

CHAPTER TEN

Maintaining and Getting the Best from Equipment

An observatory is the principal and most practical means of organising and maintaining your valuable astronomical equipment. In this chapter I will address a few of the main issues concerning optimising and maintaining observatory equipment at its best, and mention a few products and methods I have found to be particularly useful.

Collimation

Collimation is the accurate adjustment of your telescope optics. No telescope will perform at its best without perfect collimation, though errors will be most noticeable at the highest magnifications or image scales, and will always be more noticeable with shorter-focus instruments. I do know distinguished observers who hardly ever collimate their telescopes. If they work in fields such as astrometry, photometry and survey astronomy, it may hardly be necessary. For all types of imaging, and studies where high resolution is required, such as double-star work, it is crucial, however. A well-collimated telescope is also more satisfactory to look through.

Refractors

Small refractors are almost never provided with a means of collimation: they are set permanently on a lathe when they are made. They will be either right or wrong, with nothing you can do about them. Some large and short-focus refractors are provided with adjustments on the objective cell. The collimation of a refractor is only half as critical as that of a reflector of the same focal length, because of the basic laws of ray optics. If the angle of a refractive surface is changed, the ray emerging from the other side will be deflected by the same angle, whereas if the angle of a mirror is changed, the emergent ray will be deflected through twice the angle. This is also why lens surfaces only need figuring to half the accuracy of mirrors, to perform as well.

The essence of refractor collimation is that the objective needs to be square on to the axis of the tube. This is tested using a small light source, or better, a laser, placed at the drawtube, with a piece of glass between it and the telescope. The light is reflected off the glass surfaces, and, with adjustment of the objective cell, the reflection off the objective can be made to coincide with the reflection off the test surface.

Newtonians

Newtonian reflectors are far more difficult, particularly if they are below f6. The classical, and simplest, method is to use a drawtube stop, generally an old eyepiece with the lenses taken out, or a 35 mm film canister with the base sawn off and a small hole drilled in the lid centrally. This is used to centre the eye on the drawtube. All the reflections are brought into concentricity, firstly by positioning the drawtube, secondly by adjusting the secondary, and thirdly by adjusting the primary. For f8 reflectors, this is sufficient, and no further tools are necessary. For shorter reflectors, higher precision is necessary, and various accessories can help. The method I recommend for short-focus Newtonians is the method of the Barlowed laser. This was developed by Norwegian amateur telescope maker Nils Olof Carlin. I believe this to be the most accurate method, and my experience is that it is possible to use it to collimate an f4.8 Newtonian to absolute diffraction-limited performance, a feat that is almost impossible by any other means.

The Barlowed laser collimator is a gadget that can either be bought ready made, or produced by mating a standard laser collimator with a standard Barlow lens. If the drawtube is 31.5 mm (1.25 in.) fitting, the laser and Barlow need to have this fitting. If the drawtube is 51 mm (2 in.) fitting, an adaptor can be used, or the larger-fitting laser and Barlow can be obtained, though these are much more expensive, and do not deliver greater accuracy by this method in practice. The other accessory that is necessary is a stop with a hole in it, made from an opaque material, such as wood or plastic, that fits exactly in the lower end of the drawtube. Some experimentation will be necessary here, as the lower end of the drawtube may not have a standard eyepiece diameter. It is vital that this stop can be pushed in to the drawtube from the inner end, and that it will stay in, and not fall onto the mirrors. This stop is drilled centrally with a small hole. I made this stop for my reflector out of plywood, cut out of a sheet with a hole saw, which also produced the central hole. I then glued on an edging of felt (Fig. 10.1).

To use this method, the centre of the mirror must be marked. People tend to get a bit agitated about this, thinking it might damage the mirror, but in fact

Figure 10.1. Tools for precise Newtonian collimation: drawtube stop (film canister), laser collimator, Barlow, and drawtube inner-end stop.

it does no damage at all. This spot is, of course, completely shadowed by the secondary, and plays no part in the telescope's performance. The best object to make the spot from is a paper binder reinforcement ring. Simply measure the mirror as accurately as possible with a plastic ruler (which stands no chance of scratching it), determine the centre point by measuring diameters at right-angles to one another, and stick the binder ring down. Now, with the telescope re-assembled, it is possible to collimate.

First, use the film-canister-type stop in the top of the drawtube to get a rough alignment of the optics visually. The outline of the secondary should be centralised in the focusing tube, by moving it up and down the main tube and rotating it. If it cannot be centralised, the focuser may not be perpendicular to the tube, and may have to be collimated itself, possibly by shimming it. The reflection of the primary is then centralised in the secondary by adjusting the secondary collimation screws. Often, these three screws all bear on the secondary cell simultaneously, the mechanism is not sprung in any way, and it is necessary to keep the adjustment tight by loosening two screws whenever the other is tightened. Better secondary collimation systems avoid this by being spring-loaded. The primary collimation screws are nearly always spring-loaded.

Once this has been accomplished, the reflection of the secondary in the primary in the secondary (the central dark spot) must be centralised on the mirror-centring spot by adjusting the main mirror collimation screws. With one person, this is a tedious job, as one has to keep going from one end of the tube to the other, to find the screws, and then check their effect through the drawtube. Try to keep a note of which action, tightening or loosening, of which screw, has what effect on the central spot. This will make the process less hit-and-miss. Better is to have an assistant at the screws, to whom to give instructions. The result, seen through the stop, should resemble Fig. 10.2.

All this so far constitutes the rough collimation, which will be adequate in itself for an f8 instrument or above. In such a long focus instrument, the accuracy of

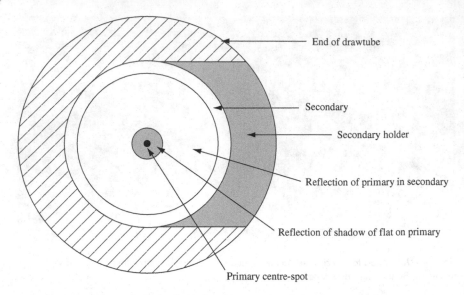

End of drawtube

Secondary

Secondary holder

Reflection of primary in secondary

Reflection of shadow of flat on primary

Primary centre-spot

Figure 10.2. The view down the drawtube of a correctly-collimated Newtonian.

collimation attained at this stage will most likely exceed the wavefront accuracy of the mirror. That is one of the beauties of long-focus Newtonians (their large range of viewing height is not).

Precise collimation is begun by removing all stops from the drawtube, and placing in it the laser, without the Barlow. This is best done with the drawtube vertical (if the tube is rotatable at all) to allow the laser to rest on the top of the drawtube without being biased to one side by a tightening of the set-screw. If this cannot be arranged, however, it is not a disaster. Check that the laser is itself collimated in its housing by rotating it in the drawtube. The spot made by the beam on the primary should not move significantly. If it does, it may be possible to collimate the laser itself by adjusting hex-head screws around its tube. When this is has been done, the beam should be centred on the centre-mark of the mirror, looking down from the open end of the telescope, by adjustment of the secondary collimation screws. The beam can be seen easily when it strikes the centre mark, but not when it reflects off a clean mirror.

The clever part comes next. The collimator is removed, and inserted into the Barlow barrel, which is then inserted into the drawtube. The "inner end of drawtube stop" or faceplate is then inserted from the inside of the telescope. The laser beam shines through the Barlow and is made to diverge. Part of it passes through the hole in the stop (only a small part, so conditions might need to be quite dark for it to be visible), is reflected off the secondary, the primary, and the secondary again, and returns to the face place, where a shadow of the primary mirror centre mark is cast. Now you will see the purpose of using, ideally, a binder ring as the centre mark. The shadow this casts is itself a ring, which may easily be centred on the hole in the face-plate by adjustment of the primary collimation screws (Fig. 10.3). The primary is then perfectly collimated, assuming the primary has corresponding physical and optical centres. The fact

Figure 10.3. Looking into the side of the telescope tube from the front, this shows the ring-shaped shadow of the centre mark produced by the laser and Barlow almost centred on the drawtube faceplate.

that it may not may mean that a slightly different result would be obtained in star-collimation.

Other methods of using a laser collimator are subject to significant error in determining the primary collimation (which is the critical part of collimation) because they are badly affected by the laser being not centrally directed down the optical axis, due to imperfectly fitting tubes, focuser slack, poor focuser collimation, and inexact determination of the primary mirror centre leading to imperfect secondary collimation. In these methods, a small error in secondary collimation, which is, in itself, not very significant, leads to a bigger error in primary collimation, which is. The method of the Barlowed laser avoids all this, through the way it creates a shadow of the centre spot, which demonstrates the primary collimation independently of other adjustments.

After laser collimation, the mirrors should be viewed through the eye-stop in the drawtube again (removing the stop from the bottom of the drawtube). They should still be almost perfectly symmetrical. If the laser procedure has resulted in the secondary edge being displaced from the primary reflection, it means the secondary is in the wrong place, laterally, in the tube. It should be moved by translation, not by re-angling it, until the reflections are again centred. The whole laser procedure should then be repeated. It is sometimes stated that, following laser or star collimation, if the collimation is then visually tested by looking down the drawtube, and residual asymmetry is still seen, this does not matter. This is wrong. Perfect collimation will satisfy all the tests of collimation, simple and advanced, simultaneously. For optimum results it is necessary re-adjust until perfect.

It may be necessary to adjust the position of the secondary in short focus Newtonians by bending the secondary supports (the spider). This is because spiders often do not take account of the *short-focus Newtonian secondary offset*. The meaning of this is that, from the point of view of the top of the telescope, in reflectors shorter than f7, there needs to be a noticeable non-centrality of the secondary in the main tube, seen looking towards the primary. The secondary is displaced away from the drawtube, as shown in Fig. 10.4. The view through the film-canister stop in the drawtube, however, *must be symmetrical*. Many books, including successive editions of *Norton's Star Atlas*, have this wrong. Getting the symmetry right in the drawtube does give you this offset automatically, and the Barlowed laser helps, but you do have to push the secondary to one side of the tube quite noticeably in f4 to f5 telescopes to get all the collimation criteria satisfied simultaneously.

Cassegrains

Cassegrains vary in the types of collimation they allow. Some are only collimatable by adjusting the secondary, like SCTs, but the best systems also allow adjustment of the primary and the focuser. Broadly, the result that should be aimed for is

Figure 10.4. Looking exactly down the optical axis of a precisely collimated f4.8 Newtonian (secondary in line with its reflection), the offset of the secondary away from the focuser is clear.

perfect concentricity of the reflections as seen through an eye-stop (such as the film canister) placed in the drawtube. The focuser should be adjusted first, if possible, then the secondary, and then the primary. A standard laser collimator can be used to collimate the secondary, but not to collimate the primary, because the beam cannot strike the centre of the primary, since it is perforated.

To use a standard laser, first check, by measurement, that the secondary is physically centred in the tube. Place the collimator in the drawtube, make sure it is collimated in its own tube, and then square the focuser on the optical axis so that the laser beam strikes the centre of the secondary. This can be judged by looking at the reflection of the secondary in the primary, by looking in from the top of the tube (take care, as always, not to look directly into the laser). The spot will be easiest to see if the surroundings are dark and the mirrors are not too clean. Once this has been accomplished, the secondary collimation screws should be adjusted (and they are nearly always non-sprung on Cassegrains, so one always has to be tightened when another is loosened) so that the beam returns centrally to the target built into the collimator. Most collimators feature a cut-away area of the cylindrical case, allowing this target to be seen from the side. The secondary is then collimated.

The primary is normally collimated by sighting the central dark spot through the stop and centring by visual estimate, followed by star-collimation. However, a different type of laser collimator is also available, called a holographic laser collimator, which projects a grid pattern of light, rather than a simple beam, at the mirrors. This can be used, with a white screen fixed some distance in front of the telescope, at night, to project this pattern on to the screen. The primary collimation screws are then adjusted until the projected pattern is symmetrical. I have not tried this method.

Schmidt-Cassegrains and Other Catadioptrics

These telescopes generally only have one user-adjustment allowed by the manufacturer, the secondary collimation. The primary is fixed to a mechanism that moves up and down the baffle tube, to focus, and cannot be adjusted. The secondary usually uses three hex-head or Phillips-headed screws, but these can be replaced by third-party knobs (such as Bob's Knobs), which are easier to use. Rough collimation is achieved as in other Cassegrains, by using an eye-stop in the drawtube and centring the circular dark reflection of the secondary. Precise collimation is done using a star. Alternatively, the holographic collimator can be employed.

Star Collimation

This is the final stage, and only complete test, of collimation for all telescope types. In practice it is rarely carried out on refractors, but frequently on other telescopes. It is highly desirable that the telescope is accurately driven, but if not, the test star should be Polaris, in the northern hemisphere, because it doesn't move very much.[1] The test star, of second magnitude, or third for large apertures (above 30 cm or 12 in.), should be centred exactly in a medium-power eyepiece (giving about 200×), using cross-hairs. If you do not have a cross-hair or reticle eyepiece, one can be made by screwing an inexpensive reticle attachment into many ordinary eyepieces. When centred, the star should be slightly de-focused, observing the appearance on both sides of focus. On one side, providing seeing is not too bad, and the telescope is thermally equilibrated with its surroundings, the pattern of diffraction rings will be clearly seen. Note, this is not the "doughnut" shape with a dark centre that is observed in telescopes with secondary obstructions when far from focus. This is the stage of defocusing just before the doughnut becomes obvious.

Further from focus, a doughnut with an asymmetrical hole will reveal bad miscollimation of the primary, and a non-circular doughnut will show that the secondary is not positioned correctly, but for precise collimation, the doughnut should be not quite visible. What should be visible is a pattern of bright and dark rings as in Fig. 10.5. The primary collimation should be altered to make this pattern symmetrical. The difficulty is that altering the collimation shifts the star in the field, even with perfect tracking, and also knocks out the alignment of any finders. So if you make more than a tiny alteration at one time, you risk losing the star.

The basic rule is that you have to find the adjustment that moves the star in the direction of the thickest side of the diffraction pattern. Then you re-centre the star in the field using the telescope's slow motion controls, or slow slew rate. The asymmetry in the pattern should then have diminished (unless you have gone too far). For a telescope that is already quite well collimated, by the methods described above, the amount of turn of a primary screw required to perfect it is likely to be less than one quarter. Again, it is sensible to keep a note

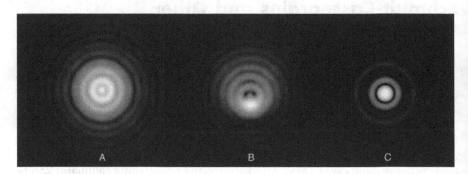

Figure 10.5. The diffraction pattern A is that given by a 20% obstructed, correctly collimated telescope, 0.8 waves out of focus. The miscollimation shown by B would be corrected by an adjustment to move the pattern upwards. C is the pattern is given by the same telescope, collimated and in focus.

---

[1]In the southern hemisphere, any one of several second-magnitude stars quite near the pole could be chosen, but they are not as good as Polaris. The southern polar star, Sigma Octani, is too faint to use as a test star for smaller telescopes.

of what adjustment you have made to which screw (perhaps number them), and what effect that has had as seen through the eyepiece. You will then get a feel for what is happening, and the process will become easier. These notes will be useful on the next occasion, if collimation is performed with the telescope in a known relationship to the sky – either pointing at Polaris, or pointing south, on the east or west side of the mounting. The rotatability of Newtonian tubes adds a further complication. A note will also need to be kept of the orientation of the Newtonian focuser.

Star collimation is easiest with a crosshair reticle eyepiece, to allow dependable centring of the star. It is ideally carried out in a series of stages, going from a medium power eyepiece to a very high power one. The very high power stage, however, say 600×–800×, will only be performable on the nights of best seeing. In this stage, only the very slightest de-focusing of the image will be required to see the pattern. On nights of very poor seeing, no star collimation will be possible, as the diffraction pattern will be too confused.

The final confirmation of collimation is to slowly focus in and out, passing through focus of the star seen at high, or, if seeing permits, very high, magnification. The pattern of light should collapse and concentrate uniformly into one circular patch (it will be quite a large patch seen at high magnification), and expand uniformly on the other side of focus. The patch should not be elongated (which implies astigmatism in the optics) or triangular (which implies usually that

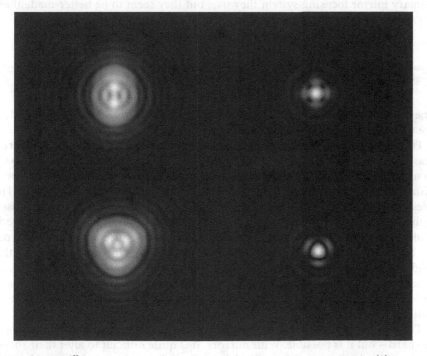

Figure 10.6. Diffraction patterns indicating optical distortions: astigmatism (top) and three-way pinching (bottom), for a 20% obstructed telescope 0.8 waves from focus (left), and in focus (right) (Images generated using Cor Berrevoets' *Aberrator* software).

the objective is being held too tightly in a three-point cell) or some other shape (which could be due to inadequate support of a mirror by its cell). Some of these possible patterns are shown in Fig. 10.6. Very close to focus, most telescopes show some small element of astigmatism, due to the action of gravity on the optics. At perfect focus, under good conditions, and at high enough magnification, faint diffraction rings will be visible (the Airy diffraction pattern, produced by any telescope). In very obstructed systems like SCTs, the first diffraction ring is in fact quite bright at perfect focus. These rings will be symmetrical if the telescope is in perfect collimation, but they are not easy to see under average conditions, particularly in refractors.

Those interested in high-resolution imaging, or splitting close double stars, will probably find that they need to re-collimate whenever the telescope is moved to point in a significantly different direction. Most telescopes require re-collimation if they are taken to the opposite side of a German mount (normalised), as the tube is then put the other way up. Some require re-collimation for more modest shifts. It is best to plan one's observation, so far as possible, so that all the objects to be observed are near the meridian when they are observed, minimising these shifts, and also maximising the objects' altitude. But this cannot be applied to objects near the celestial pole, or to those that appear temporarily, such as comets, and may have to be observed in the northern sky (for a northern hemisphere observer), or wherever they happen to be. Schmidt-Cassegrain telescopes used to be particularly bad in respect of collimation shift, because of the movable primary mirror focusing system they use, but they seem to be better-made these days, with less "mirror flop".

Cleaning Optics

The subject of cleaning optics causes a lot of anxiety and misunderstanding. Here are a few ground-rules to keep in mind.

Firstly, in general, you can clean all your optical surfaces yourself, if you are a reasonably careful and competent person. There is no need to send things back to a manufacturer for cleaning. However, some people may be nervous about dismantling a complicated telescope and putting it back together again. The telescope in question is likely to be an SCT or other catadioptric. In such cases, if the innards get noticeably dirty, sending it back to the maker, or to a good dealer, might be the preferred course. For those living in places far-flung from the centres of telescope manufacture, however (which includes most people), this might be judged impractical.

Secondly, cleaning optics does not require special or expensive equipment. Typical equipment would be distilled water, industrial alcohol, or a proprietary optical cleaning liquid, and tissues, or cotton cloths, or cotton buds.

Thirdly, it is not as easy as is often thought to damage optics by cleaning them too hard – but it is possible. With mirrors, it is quite difficult to alter their shape, because they are protected by their coating of aluminium and the oxides that form on top of that, together with any overcoating that has been deliberately applied. Basically, so long as you avoid scratching them, by carefully removing all dust and grit with a soft brush, or air blower, before cleaning, and so long as the coating remains intact, you cannot alter their figure. You should still be gentle, to avoid removing coatings that may have flaws, due to ageing.

Lenses are more sensitive things, as nothing much protects them, particularly if they are not anti-reflection coated (most are these days). Broadly speaking, the closer to the eye, or camera, that a lens lies in the optical train, the less critical is its figure. No harm comes from rubbing or polishing eyepiece lenses, nor, probably, Barlows, though they are less prone to getting dirty in the first place.

Eyepiece lenses can be cleaned with a small piece of cloth, or cotton buds, and alcohol, distilled water, a mixture of these, or an optical spray. They may be cleaned in-situ, or, for a better result, the eyepiece may be disassembled. I don't tend to do this very often, as it is tedious. You must note carefully the order in which the components were assembled, and the way round the plane, convex and concave surfaces of the lenses were, in order to put them properly back together. Lenses are generally held in with slotted, threaded rings that would require a special tool, which you probably will not have, to undo them and do them up again. You can usually make do with flat screwdriver blades, but there is a danger of the blade slipping, and causing damage, and it can be very difficult to get some eyepieces apart. In general it is only the outer surfaces of the outer lenses that need cleaning, and these can cleaned well enough without disassembly. First, remove dust with a photographers' blower brush, or similar, or a can of pressurised air, and then polish gently with a wet cotton bud, followed with dry ones.

Objective lenses and corrector plates are the most sensitive optics, and those which need to be treated with the greatest respect. The golden rule is to apply minimum pressure, and not to rub, and *not to polish in circles*. If you polish in circles, you will re-figure the optic. The cleaning strokes should be small parts of an arc, with a lifting motion at the end, to remove dirt and debris with the cloth or tissue. Change to a new part of the cloth, or use a new tissue, frequently, possibly every stroke.

Many people recommend using a clean cotton cloth, but these are not necessarily to hand, and anyway, I find they tend to deposit fibres on the surface being cleaned. Ordinary facial tissues are quite adequate, but do not use toilet tissue, as this contains greasy additives. Some people recommend cotton-wool buds, but, again, ensure they are free from additives (they should have no smell). "Baby" products are likely to be additive-free. Again, remove dust and grit carefully with a brush or blower before commencing with the liquid cleaning. Some people recommend using a very dilute soap-powder solution, and that may be a good idea for very dirty optics, but I prefer an ordinary commercial "spectacle lens" cleaning liquid, available from any ophthalmic opticians' shop. This is just an alcohol-water mixture, and will work fine. You may be able to save yourself money by mixing up your own solution, using distilled water, but this may not be worthwhile. Do not use tap water; this leaves deposits.

The final rule is not to be perfectionist. Do not expect to see no dust or dirt on your optics, and do not expect to be able to see no streaks, mis-colourations, or marks of any kind on them either. Optics are there for looking through, not at. Small imperfections, particularly miscoloured streaks caused by thin-layer interference on lenses, corrector plates and optical windows, have no detectable effect on performance. These streaks most commonly show up as pink stains on a bluish anti-reflection coating under obtuse reflection. Do not worry about things like this. Do not clean mirrors and lenses until they are quite obviously dirty or marked or greasy, and then do not "make a meal" of it. Just get them reasonably clean.

Try to prevent optics from becoming dirty in the first place. Always replace the caps after an observing session (unless the optics have got damp and need to dry out), and store eyepieces in bolt cases or other suitable receptacles (not cardboard boxes). Store filters in plastic boxes. Use a clean lens-brush from time to time to remove dust, and disassemble things only when unavoidable. Mirror or prism diagonals sometimes need a proper clean, and their optical surfaces are got at by unscrewing the barrels from each end of the housing, or by unscrewing a back-plate, held in with small screws at the corners, that holds the mirror or prism in. Clean in the same way as lenses.

In passing, the cleaning of CCD and CMOS chips might be mentioned, as these are a form of optical surface. To clean the relatively small chips found in webcams and astronomical CCD cameras, I find the best appliance, again, is a cotton bud. It does not need to be wetted, except in bad cases. Draw it over the chip, from one side to the other, in one sweep, lifting at the end. Larger chips require several sweeps from points along their short edge, to cover them with the bud. It is hard to tell if they are actually clean enough without using them to take an image. Dust may be visible if you examine them under good lighting with a magnifier, but significant dust and dirt is good at hiding on imaging chips, and only revealing itself onscreen at night. The best way to detect it is to connect the camera up to the computer to view the image, and to shine a light on to the chip past an obstructing opaque edge, the shadow of which is drawn across the chip. The imperfections appear on-screen as this shadow is drawn across.

Unfortunately, dust often falls back onto webcam chips when the plastic threaded adaptors are screwed back on to the cameras. Also, do not expect a new chip to be perfectly clean. They rarely are, and they all have to be cleaned from time to time. Some are protected by an optical window, which means that that will be the surface needing more frequent cleaning, with the chip only needing attention occasionally. Dropping cameras tends to make the detectors very dirty. Optical spray can be used in very small quantities, polished off with dry cotton buds.

The large chips in DSLR cameras are particularly problematic with respect to dust. Some recent cameras have self-cleaning mechanisms, but I have no idea how effective these are. DSLR cameras have to be powered-up and put into a cleaning mode in order to access the detector, which is normally concealed by the mirror and shutter. I have not found that blowing air at the detector does much good: it tends just to redistribute dust. Again, I recommend a few gentle wipes with dry cotton buds (but please do not sue me if you damage your camera). Other authorities insist that these chips must not be touched, but must be returned to the manufacturers for cleaning. For most people, this advice will be ridiculously impractical.

Adjustment, Cleaning and Lubrication of Mechanics

The main areas to be covered here are focusers and mountings. Rack-and-pinion focusers can generally be adjusted with the screws that bear on the plate over the pinion (the small cog wheel). They are often supplied done up too tight, in the hope they will wear in. If the drawtube seems too tight, also ensure that there is not a clamping set-screw that is done up. If too loose, focusers will be wobbly, and may rack out or in of their own accord when heavy equipment is attached to them. Poor-quality focusers are often very "sloppy" anyway. Experiment with the various adjusting screws. Very high quality focusers, such as the Crayford and micro-adjust types, are becoming increasingly common, and are infinitely better than the things we had to cope with in the past.

Simple rack-and-pinion focusers are easily disassembled for cleaning, by completely undoing the plate over the pinion, removing the pinion with its knobs, and pulling out the drawtube. This is necessary if the focuser is very stiff, and cannot be eased by adjustment. Clean using alcohol, white spirit or methylated spirit, and re-lubricate. Vaseline petroleum jelly is recommended for lubricating any parts that come close to optical surfaces. It is also very useful for thinly coating the threads of aluminium camera and telescope adaptors and filters. These otherwise can often become jammed in place for no particular reason. The benefit of Vaseline is that it does not tend to "creep" on to the optical surfaces.

For mountings, a higher quality of grease is required. Molybdenum grease, which looks dark, or Teflon grease, which looks white, are possibilities. The latter is stocked in cycle shops. The grease should not be too thick. Modern mountings have mechanisms which are not readily accessible. The worms and wheels are usually enclosed to keep them clean. This is fine, if they have been well-adjusted and appropriately lubricated in the first place. Unfortunately, the cheaper models all too often have not been. It is not possible here to go into the details of servicing mountings as there are so many variants, but it generally can be done by the user as they are not that complicated (though you may invalidate your warranty if you still have one). Detailed instructions for some of the most common types have been placed on the web by enthusiasts. The basic rules would be to work in a clean and tidy environment, to be careful not to lose or overlook any small parts, and to lay parts out in the order in which they were assembled, noting which way round they went. The objective will generally be to ensure smooth and free rotation of all the shafts and gears while minimising play and backlash. Minimising backlash in gears is something of an art, and many attempts may be required to get it right.

The lubricants used in the cheaper mounts are often far too thick and viscous, and should be cleaned out completely and replaced. Older mounts are generally simpler, and have the gears unenclosed, making them easy to clean (and easy to get dirty again). They are easier to adjust to minimise backlash, because it can be seen precisely what is going on in practice, but often they have severe backlash problems which cannot be solved, as they are due to bad manufacture or design.

One comes back to the point that investment in a high-quality mounting will always pay dividends. Mountings are equally as important as optics, and should have at least as much spent on them, for optimum observational results.

Preventing Dew

As has been mentioned before, dew on optics is a major headache for most observers. Dew caps or shields generally only delay dewing for between 30 and 60 minutes. Newtonian and Cassegrain reflectors in solid-wall are the least susceptible telescopes to dewing, as the objective is well-shielded, but the secondaries often dew up, particularly on larger telescopes, as of course do the eyepieces and the refracting finders, at both ends. The problem is most severe with large-aperture catadioptrics, where the large area of the corrector plate takes a lot of warming to keep dew at bay.

Occasional blasts with a hairdryer are the response of many observers to dewing, but this is not ideal. For one thing, this is likely to overheat a small part of the optics, and cause optical distortions. For another, mains hairdryers are dangerous in an observatory or outside. Attempting to clear a Newtonian or Cassegrain secondary with a hot air blast tends to blow the moisture down to the objective, where it condenses again. For emergency use, a 12 V hairdryer or heater is a safer and gentler remedy. This is good for clearing eyepieces and diagonals. If a cap is always placed on the eyepiece when it is not being looked through, dewing here can be minimised, but it is difficult to remember to do so. If all the accessories to be used in a night, eyepieces, filters, and diagonals, can be warmed up in advance of the observing session and kept in an insulated or warmed box, this can limit problems, but somehow, on the worst nights, it never prevents them entirely.

Refractors generally have a reasonable dew-shield built-in, and the addition of a 12 V dew-strip should completely prevent dewing, while not causing any noticeable seeing or optical distortions. Catadioptric telescopes are, on average, larger, and while the power of manufactured dew-strips generally increases in proportion to their length, which is proportional to the aperture, the surface area and volume of glass to be warmed increases in proportion to the square of the aperture. Consequently, dew strips have a difficult task warming the corrector plates of catadioptrics and large windowed telescopes sufficiently, and the result is often dew in the middle of the plate, with clear glass around the edge. Most users of these telescopes find they need both a dew-strip and a long dew-shield to always keep dew at bay. The dew shield contains more of the heat generated by the strip near to the corrector, greatly increasing efficiency by creating a "micro-climate". Another advantage of dew shields is that, if well blacked inside, they reduce the stray light, such as moonlight or artificial light, entering the telescope, and thus increase contrast. Good ones have a velvety black flocking inside, which also absorbs condensation and prevents it from running down onto the telescope.

The dew-shields sold for catadioptrics are of the flexible and rigid varieties. In my experience, the flexible ones do not work very well. They are either made of some kind of fabric, or thin plastic, and are easily folded for stowage. However, this is not necessarily what the observatory-owner wants. He maybe wants a permanent solution, whereby the telescope is always fitted up in such a way that dust is kept off the optics when not in use, dew never forms on them, and nothing has to be taken on or off or fitted to the telescope when starting observing and closing down, other than a dust-cap. These requirements do not seem to have occurred to the major manufacturers, who continue not to supply dew shields as standard, and also not to make available rigid dew-shields for the larger sizes. Where they are made, those I have seen all fit in such a way that the telescope cap cannot be fitted securely over them. So the telescope must stored without the shield, then uncapped at the beginning of the session, the dew-shield must be fitted, which is usually an awkward process, the dew strip is usually then fitted over or next to the shield, and then all this needs to be put into reverse at closing-down time. It is not an integrated solution. And the flexible shields offered for the largest apertures are even more of a pain to use. There are small accessory manufacturers who offer rigid shields for larger apertures, but outside North America their wares are difficult and expensive to obtain.

As an aside, there is a general problem of telescope manufacturers not catering to observatory-owners very well, but tending only to direct their products at mobile astronomers. The 12 V cigarette-plug issue mentioned earlier is one example. Another is the impossibility of buying most mountings, apart from the very largest and most expensive models, without a tripod, which the observatory owner generally will not want. It would be useful if manufacturers would research better the wants of the growing number of small observatory owners. With anti-dew equipment, all that is needed is for SCT owners to be able to be in the same position as refractor owners, with properly thought-out dew-shields built-in to the telescopes, which could be retractable or removable for mobile users. It would be more efficient if dew-heaters could be inserted into the insides of telescopes, rather than wrapped round them, but this would be difficult to retrofit, and again needs to be built-in by the manufacturers. Meade have recently started to produce telescopes that have this, in their RCX400 range.

Flexible dew shields for large scopes can be made, much more cheaply than they can be bought, out of the foam mat material sold in outdoor shops. I tried one made out of lacquered card, but it didn't last very long. Rigid dew shields are far more satisfactory, but not easy to make for larger telescopes, though Martin Mobberley managed it, as mentioned in Chapter 9.

Dew-controllers are boxes that plug into a 12 V source, usually using the terrible cigarette-lighter plugs, and allow one or more dew strips to be plugged into them, usually using phono plugs. They may have variable output for the different channels, so more power can be provided to a large dew strip on a corrector plate, less for one round a finder, and less again for one on an eyepiece. The circuitry is actually an astable multivibrator, which modulates the power to the strips by feeding them a square waveform, so they are only on for a proportion of the time. This is more efficient than using a rheostat (variable resistance). The circuitry is quite easy to construct for anyone with electronics skills. The idea is to feed each of the strips only so much power as is needed to raise the temperature of the optic by a fraction of a degree Celsius, which is sufficient to prevent dew. Minimal heating is desirable to avoid distorting the shape of the optic or causing turbulence, but I have never seen any untoward effects of dew-heating in practice, and I have not used dew controllers, but merely wired-up a splitter cable for my multiple strips, using line phono sockets, available from electronics suppliers, and a terminal block.

Polar-Alignment for Equatorials

Modern German equatorial mountings often come with a polar scope, a small telescope running through the polar axis that can be used to sight the star-field around the pole. This is convenient for portable use, but unnecessary in an observatory if the mount is never to be moved. The best method of aligning a fixed mounting (this applies to any type of equatorial), and the only really precise one, is the drift method. This is not very practical for portable instruments, as it is time-consuming. With an observatory, the alignment can be refined by the drift method over many nights, and can be made, in principle, perfect.

Mounts vary in how well and precisely the altitude and azimuth adjustment mechanisms work, and these make attaining perfect alignment easier or more difficult. Old mounts often have only a single bolt running though the base casting, connecting it to the polar casting, and when this is slackened off, the whole mounting and telescope is uncontrollably (and possibly dangerously) free in azimuth, and may collapse. Precise adjustment of such a mounting in altitude is very hard. Also, there generally is no provision for azimuth adjustment on such mountings at all, so the whole telescope has to be yanked round on the pier top (also potentially dangerous), possibly using a crow-bar for leverage.

Modern mountings are mostly far better, and usually have screw-adjusters above and below the polar assembly, that will push it up or down in altitude (when one is tightened, the other must be slackened off), and a similar push-push screw assembly to fine tune the azimuth. To use this, the azimuth setting must first have been determined correctly, to within a few degrees, before screwing the mount-base to the pier-top.

Since this initial azimuth setting is probably going to be done during the day, it can be done with a magnetic compass, allowing for magnetic variation, or, more accurately, by observing the transit of the Sun. The Sun transits, not at local mid-day, but at the times given in Table 10.1 for each date in the year (this difference between local mid-day and sun transit is known as the Equation of Time). The first requirement is to determine the difference between local time and the standard time of the time-zone in which the observer is situated. For each degree in longitude, the change is 4 minutes of time, getting earlier towards the west. Then apply the Equation of Time correction from the table. Attach a gnomon to the mounting (something like a ruler, to cast a straight shadow), or use the tube-rings, and, at the determined time, assuming you have the co-operation of the weather, turn the mount to bring the shadows into line, and you have your initial azimuth alignment. The *levelness* of the mount, as has been pointed out before, is not important. Errors in this are overcome by the drift method. The initial altitude setting can be made with a protractor.

Table 10.1. Local time of solar transit (the Equation of Time) throughout the year.

Jan. 0	12:03.0	10	12:01.4	19	12:06.3	27	11:43.9
5	12:05.3	15	12:00.1	24	12:06.5	Nov. 1	11:43.6
10	12:07.4	20	11:59.0	29	12:06.5	6	11:43.6
15	12:09.3	25	11:58.0	Aug. 3	12:06.2	11	11:44.0
20	12:10.9	30	11:57.3	8	12:05.7	16	11:44.7
25	12:12.3	May 5	11:56.7	13	12:04.9	21	11:45.8
30	12:13.2	10	11:56.4	18	12:03.9	26	11:47.2
Feb. 4	12:13.9	15	11:56.3	23	12:02.7	Dec. 1	11.48.9
9	12:14.2	20	11:56.5	28	12:01.3	6	11.50.9
14	12:14.2	25	11:56.9	Sep. 2	11:59.8	11	11:53.1
19	12:13.9	30	11:57.5	7	11:58.1	16	11:55.5
24	12:13.3	Jun 4	11:58.2	12	11:56.4	21	11:57.9
Mar 1	12:12.4	9	11:59.2	17	11:54.6	26	12:00.4
6	12:11.3	14	12:00.2	22	11:52.8	31	12:02.8
11	12:10.1	19	12:01.3	27	11:51.1		
16	12:08.7	24	12:02.4	Oct. 2	11:49.4		
21	12:07.3	29	12:03.4	7	11:47.9		
26	12:05.8	Jul. 4	12:04.3	12	11:46.6		
31	12:04.3	9	12:05.2	17	11:45.4		
Apr. 5	12:02.8	14	12:05.8	22	11:44.5		

The drift method is used at night, and can be performed with an eyepiece, or a CCD camera and computer. An eyepiece needs to be equipped with a crosshair reticle. Locate a reasonably bright star near the meridian, and, preferably, somewhere near the celestial equator (0° dec.). Let it drift, or use the E or W button on the controller, to determine E and W in the eyepiece field, and rotate the eyepiece so that the crosshairs are aligned E-W. If using a CCD and computer, rotate the camera so that the E-W direction is horizontal on the monitor. Many image capture programs will produce a reticle on the screen. Use the telescope controls to place the star in the centre of the field. Allow the telescope to track for a few minutes. The star will either have drifted north or south in the field. If you are not sure which is which in the field, apply a bit of pressure to the telescope tube to bias it in the direction of the pole, and see which way this moves the star. If the star moves in the direction it has drifted, it has drifted south. If the opposite, it has drifted north. (This applies to the northern hemisphere.)

If the star has drifted northwards in the field, this means that the polar axis of the telescope (running upwards, in the northern hemisphere), is pointing too far west, and therefore, as seen from above, the mount needs to be rotated slightly clockwise. If the star has drifted south, the polar axis is too far east, and the mount, as seen from above, needs to be rotated anti-clockwise.

Having done this, choose another fairly bright star, this time one due west, that is, at right angles to the direction of the polar axis, and 20°–40° above the horizon. If this cannot be done, because that horizon is blocked, choose a star of similar altitude due east. Place it on the crosshairs, or on-screen reticle, which will still be aligned with the co-ordinate directions (unless a Newtonian tube has been rotated), and observe its drift over a few minutes. Observe whether it drifts north or south, if necessary confirming again which direction is north, and which

Table 10.2. Rules for the drift method.

	Northern Hemisphere	Southern Hemisphere
Star on the meridian	Drifts north: rotate mount clockwise	Drifts north: rotate mount clockwise
	Drifts south: rotate mount anti-clockwise	Drifts south: rotate mount anti-clockwise
Star in the west	Drifts north: raise polar axis	Drifts north: lower polar axis
	Drifts south: lower polar axis	Drifts south: raise polar axis
Star in the east	Drifts north: lower polar axis	Drifts north: raise polar axis
	Drifts south: raise polar axis	Drifts south: lower polar axis

south, in the eyepiece. The direction of drift this time shows whether the polar axis needs to be pointed higher or lower. If a star to the west drifts north, the polar axis is too low, and must be raised, in the northern hemisphere. If the star drifts south, the polar axis must be lowered. A full table of the use of the drift method for both hemispheres is given in Table 10.2.

The method should be used iteratively. So having made the azimuth and altitude adjustments, go back to the meridian again with a new star and test the error in azimuth again, and so on, over many cycles of either increasing duration or higher magnification, and, assuming it is working correctly, increasingly fine adjustments, so you do not overshoot. When you reach the setting at which there is no detectable drift in a medium-power eyepiece (about 80×) for 15 minutes, this will be good enough for virtually all purposes. Drifts due to refraction and instrument flexure will negate higher precision. This process may be spread over many nights – so only with an observatory is this optimal polar alignment possible. When it is completely satisfactory, additional fixings on some observatory mounts are tightened up to lock the alignment permanently. Sometimes, tightening a mount down to fix the azimuth setting can slightly affect the altitude setting, so this should checked again as the bolts are progressively tightened down.

Balance, Accessories and Extra Telescopes

On an altazimuth mounting, the telescope must be balanced about the altitude bearings. Either it needs to be possible to slide the telescope through tube rings, which are then tightened-up, to fix the balance point, or an extra counterweight assembly must be attached to the tube, which provides a means of altering the centre of gravity, either by adding or removing weight at a fixed distance from the altitude axis, or by moving a fixed weight up and down the tube on a rail mechanism, with provision for locking. Reflectors and catadioptrics tend to balance rather close to the main mirror end, refractors much closer to the middle. The shorter the primary focal ratio, the more asymmetrical the balance tends to become. The best solution may be found to be a combination of a coarse adjustment of the tube up and down the mounting, done only once, with fine adjustment using a counterweight rail, done frequently to compensate for changes in the equipment fitted. Owners of SCTs screwed to fork mounts will have no means of adjusting the tube up and down the mounting, and also in many other systems, changing the balance of the tube by moving it is not easy; therefore, in these cases, small counterweights on rails are the only option.

Such rail systems are rather expensive to buy, but not difficult to fabricate, with a bit of basic ingenuity. Fig. 10.7 shows a design that I have developed. The rail is a tube or rod of metal, attached either to the tube or to the mounting plate or dovetail bar using some kind of bracket. The moving part consists of a disk or other shape of aluminium, plastic, or wood, at least 12 mm (0.5 in.) thick (I used aluminium). A hole is drilled in this to allow it to slide on the rail. The weight is a small, say 0.5 or 1 kg (1 or 2 lb.) barbell weight, as can be had from fitness shops. A machine screw or bolt of appropriate length (some experimentation will be required to get this length right) runs through the barbell weight and is clamped tight to it using a nut and two large washers (plate washers of the type used in timber construction). The running piece, the disk or other shape of aluminium, plastic or wood, needs to be tapped to accept the screw or bolt. If it is plastic or wood, this may be possible without using a tap, by just forcing the screw or bolt into a hole slightly too small for it. The machine screw runs through the side of the running-piece and into the central bore just far enough to contact the rail. The device works by twisting the disk weight to clamp or unclamp the running piece on to the rod. When undone, the weight assembly can be slid along the bar until the balance point is found, and it can then be done up. I find this device, if carefully made, works very well.

With an equatorial mounting, longitudinal balance of the tube needs to be established about the declination axis, as it is about the altitude axis in an altazimuth mounting. Additionally, with the German mounting, appropriate counterweighting of the whole telescope assembly, with all its accessories and longitudinal counterweights, if any, needs to be provided at the far end of the dec. axis. With a German mounting, one can find oneself getting into a lot of complication, and adding more and more weight to the system, trying to counterweight counterweights.

With a Newtonian on a German mounting, there is an extra problem, as the tube assembly is not naturally balanced around its longitudinal (or main optical) axis due to the focuser and whatever equipment is attached to that, as well as to finders. Additionally, it is, as has been mentioned before, highly desirable to be able to rotate the whole tube in its rings in order to get the focuser or finder to a convenient position. This rotation will alter the balance of the system about the polar axis, as the weight of the focuser etc. either opposes the dec. counterweights, or lies more towards them (Fig. 10.8). This should be countered by making the tube at least approximately balanced radially, by adding weight opposite the focuser and finder. This might be the best position for the sliding weight assembly. Perfect radial balance is not required, nor is it necessary to make a lot of complex calculations to establish how to achieve it. If the mount moves so freely that the slightest change of weight distribution takes it out of balance enough to make it drift, then the clutches probably need tightening. A telescope should not need *perfect* balancing.

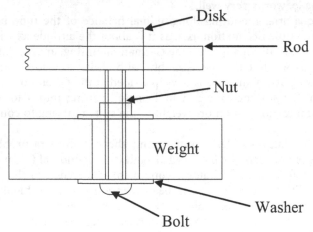

Figure 10.7. Simple home-made movable balance weight assembly.

It must be said that adding heavy cameras, filter wheels, and so on is a particular problem with Newtonian systems, as such equipment can be quite far from the longitudinal axis of the telescope, and can have a large effect on balance. Refractors and Cassegrain-configuration telescopes are easier in this respect; they need not be rotatable, and only simple longitudinal balance, and

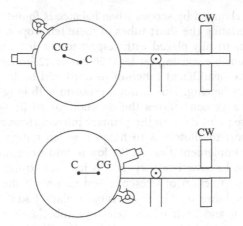

Figure 10.8. Change of balance as a Newtonian tube is rotated (C is the centre of the tube, CG is the centre of gravity, CW is the dec. counterweight).

then dec. balance, need to be established. Short Cassegrain and catadioptric tube assemblies are commonly balanced using a dovetail saddle and dovetail bar arrangement, where the tube, or the tube rings, are attached to a bar which fits into a wide groove in a wider piece of metal, the dovetail recess or saddle, which may be an intrinsic part of the mount, or a component bolted to it (Fig. 10.9). The dovetail bar can slide up and down in the saddle to attain balance about the

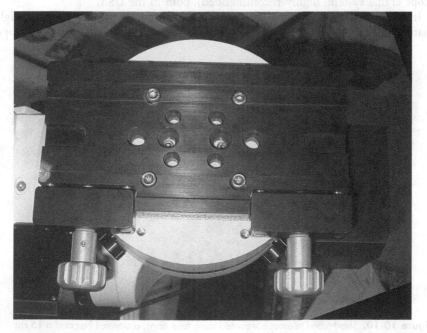

Figure 10.9. Dovetail saddle of the Losmandy/Astro-Physics/Robin Casady type.

dec. axis, and it is clamped by screws when balance is found. These systems are useful in accommodating the short tubes of these telescopes, which, if mounted in tube rings symmetrically placed with respect to the dec. axis, and bolted to a saddle-plate at a sensible spacing (at least 30 cm or 12 in. for a moderate-sized telescope), will have insufficient range of forward and backward movement, so preventing proper balancing. This is not a problem with long telescopes.

The dovetail arrangement allows the dovetail bar to be supported near the back, at the balance point close to the primary mirror. However, the saddle part has to be quite short to allow this to happen, which mitigates against a really firm linkage. In its implementation on the low to mid-price mounts, the dovetail bar system tends to be unsatisfactory as it is just not strong enough. It creates a weak point at the junction of telescope and mount. If the bar is substantial enough, it can work, however. The result is particularly bad if a thin bar is bolted directly to the front and back of the telescope tube, as is common with SCT systems, as this allows a lot of flexure and vibration. Strong tube rings will damp this vibration out.

A good compromise for mounting heavy, short-tube instruments is a pair of tube rings attached to a substantial dovetail bar interlocking with a reasonably long dovetail saddle (at least 15 cm or 6 in. long), as shown in Fig. 10.10. The length of the saddle will limit the range of adjustment available, so an additional adjustment potential using sliding weights is very desirable, particularly if heavy cameras are to be added. A problem with this recommendation is that it is rather difficult to get adequate tube rings to fit large telescopes. The major manufacturers do not make them (with the exception of Takahashi). Satisfactory products are made by Parallax Instruments and Andy Homeyer (his semi-circular scope cradle system, highly-recommended), both in the USA.

It always much easier to make a balance adjustment by moving a small weight than it is by loosening the dovetail clamps and shifting the whole telescope,

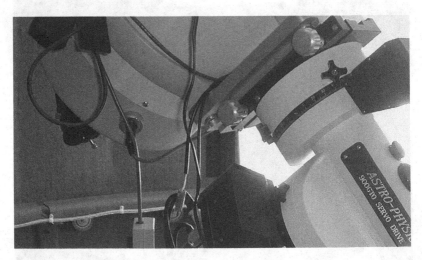

Figure 10.10. Short-tubed telescope mounted using tube rings, a dovetail bar and a 15 cm (6 in.) dovetail saddle.

which is hard work, risky, difficult to do precisely, and risks knocking out the collimation or the finder alignment. Hence I recommend moving counterweight systems for all substantial telescopes, whether supplied with dovetails or not. If maximum rigidity is desired, dovetails are best avoided, and the tube rings should be bolted to a rigid mounting plate, bolted directly to the mount. The fixings can be drilled so the mounting is asymmetrical, bringing the line of the dec. axis through the balance point near the main mirror. Precise adjustments are then made using small weights.

The counterweighting on the dec. axis is best in the form of several smaller weights rather than one large one, or one large one and one small one, so that fine adjustments to compensate for changes in equipment can be performed without moving most of the weight. Older telescopes often use collars round the shaft on either side of the weight, secured with Allen screws, and these are difficult to fiddle with at night. The better system, which new mountings normally have, consists of a hand-knob in the side of the weight itself, bearing upon a plunger which clamps the weight to the shaft. This can, however, cause a problem with wires getting caught on the knobs. Whichever system is used, it is imperative, for safety reasons, to ensure that the weights cannot possibly slip off the shaft when their fixing is loosened. Most mounts have a safety stop on the end of the counterweight shaft to prevent this from happening. If this is lacking, a hose clip or Jubilee Clip of appropriate length, tightened round the end of the shaft, can do the same job.

A telescope on a German mounting must always be assembled by placing the counterweights on the shaft first, and then securing them, and then attaching the optical tube. When disassembling, the tube comes off first, then the weights. Attaching a large optical tube via a dovetail requires a lot of care. It is very easy to believe the fixing is secure, when it is not. With a heavy tube, the weight can bear on the dovetail joint in such a way as to make it immovable, even when the clamps are not done up thoroughly. The clamps may seem to be tight because of the weight bearing on them, but shifting the weight of the telescope can then make the whole thing come loose, and, in the worst case, the telescope can fall out. This actually occurred to one observer I know, and he took the picture of the aftermath Fig. 10.11. The telescope was badly damaged. The moral is to be very sure about the fixing of telescope dovetail bars in their slots. Shift the weight of the telescope about, with the mount clutches disengaged, while holding the optical tube, and re-tightening the clamps, to ensure they are really tight in all positions, before letting the mount run. Alternatively, avoid dovetail fixings.

The installation of a guidescope, or another subsidiary telescope, on the same mounting as the main instrument presents special problems of preserving balance. The guidescope is normally a small refractor, though it could be any type of telescope, and it has to be accurately alignable with the main telescope, and capable of being offset by at least half a degree from that alignment, in order to locate a guide-star. This distinguishes the guidescope from a subsidiary or secondary telescope, which does not need to be accurately alignable. The normal method of making the guidescope alignable is by using two guide-scope rings with three screws each, though there are other devices, for example the Robin Casady Tandem Guidescope Aiming Device.

Figure 10.11. A photo of the aftermath of an actual incident in which a telescope (with camera) fell out of a dovetail fixing.

Most guidescope rings have three simple screws, tipped with nylon, but this is not the best system, as it means that whenever a screw is tightened, another has to be loosened (the same problem as with many collimation systems). The better system is to have one of the screws replaced with a sprung stud, so that adjusting either of the other screws on its own changes the alignment of the telescope, while the system remains in compression. I have spring-loaded a normal guidescope ring and screw arrangement with three springs located between the rings and telescope (it is only necessary to do this on the ring you intend to adjust, normally the rear one) as shown in Fig. 10.12. I placed round, curved plastic washers (cut out of a plastic drainpipe with a holesaw) under the cut-to-length springs to prevent them from scratching the guidescope. I find this arrangement works much better than the standard guidescope ring system for obtaining precise adjustment.

Popular guidescopes at the moment are the ST80 short-tube 80 mm (3.5 in.) refractors, which are small and light, but something higher-quality might be desired in order to have the option of using the guide scope as an imaging scope in its own right. Such small imaging scopes are useful for plugging the image-

Figure 10.12. Guidescope ring modification with springs and plastic disks.

scale gap between telephoto lenses and large telescopes, a range which is needed for imaging large galaxies, and most nebulae and open clusters, on standard CCD chips. For this purpose, apochromatic refractors ("apos") in the range 75–125 mm (3–5 in.), with focal ratios about 6, are very popular and effective.

The standard way to mount the rings of the guidescope is to bolt them to the top of the main scope rings, or to a rail which is bolted to the main telescope, or to its rings, opposite the dec. axis. Higher-quality tube rings offer this facility, with a flat mounting surface at the top. However, this arrangement places the guidescope very far from the polar axis, and thus gives it a large moment, requiring counterweighting on the dec. axis disproportionate to the guidescope's weight. This will strain the mount more and lead to more flexure and less stability. It also tends to make the guidescope get in the way in a small observatory, since the most common azimuthal direction for the telescope is the meridian, and, in this position, the guidescope will lie horizontally alongside the main scope. For these reasons, I concluded that, for my main set-up, this was not the best arrangement, and I mounted my guidescope much more "in-board", towards the junction of the main telescope and the dec. axis (Fig. 10.13).

If the guidescope is off the line of the dec. axis, however, it has to be counter-weighted by another weight on the opposite side of the axis, otherwise the set-up will be unstable. It could be in a balanced equilibrium, but it will be an unstable equilibrium. Figure 10.14 illustrates this point. The centre of gravity of the system (C_2) will be placed above the line of the dec. axis in certain telescope positions, and will want to fall, i.e., to flip the whole thing over. It is, however, not necessary to add a mass equal to the guidescope's on the other side, unless the guidescope is placed at 90° to the dec. axis, which I do not recommend, as that maximises its moment and maximises the counterweighting needed and, also, the counterbalancing weight required for both on the dec. axis. Better to

Figure 10.13. Positioning of my guidescope near to the mounting, also showing my two-dimensional moveable weight balance system.

adopt the "inboard" arrangement I propose, in which a smaller counterweight opposes the guidescope, in a position that is not obstructive, bringing the centre of gravity back onto the line of the dec. axis, and in which a small amount of extra counterweight at the other end of the dec. axis brings the centre of gravity of the system back to the RA axis.

In Fig. 10.14, the guidescope could be counterbalanced by an equal mass located at G', which would bring the centre of mass of the telescope and its attachments back into line with the dec. axis, but that would be a bad plan, as it would add a lot of moment that would need to be counterbalanced on the other end of the dec. axis. The optimum arrangement, my system, is shown in the lower part of the diagram, with a small counterweight attached to the saddle-plate, on the opposite side of the dec. axis to the guidescope. The advantage of this is that, because the distance d_2 can be greater than the distance d_1, the mass of the counterweight can be much less than the mass of the guidescope, minimising the extra that has to be counterbalanced on the opposite side of the RA axis. Additionally, in my arrangement (Fig. 10.13), this weight is placed well forward of the dec. axis, contributing to the counterbalancing about the dec. axis of a heavy Cassegrain primary. In this position, the small counterweight never collides with

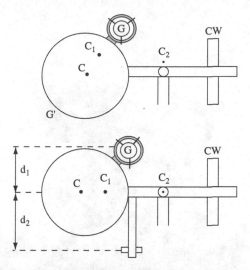

Figure 10.14. Balancing a system with a guidescope out of line with the dec. axis: C is the centre of mass of the main telescope, C_1 is the centre of mass of the main telescope plus its attachments, C_2 is the centre of mass of the whole system including dec. counterweight (CW), G is the guidescope.

the mount, even when the telescope is below the main counterweight level, on the "wrong" side of the mount.

The counterbalancing system I have constructed allows the adjustment of weights mounted on rods (the mechanics as described earlier) in two directions, along the main tube and at right-angles to it. Any combination of accessories on the main and guide telescopes can be balanced in this way, without the need to adjust any dovetail fittings. The rods on which the weights move are attached to the dovetail mounting plate using copper plumbing-type pipe clips, fixed with nuts and bolts. Because it is not attached to the telescope, this whole assembly remains in place when I swap my Cassegrain for an SCT, as does the guidescope, attached to the lower part of the tube rings.

The guidescope, in this arrangement, will restrict the motion of the telescope if it goes below the level of the main counterweight when the guidescope is below the dec. axis, as then the guidescope is potentially in collision with the mounting. Hence this is a position I do not use. However, those with a large observatory and over-specified mounting may prefer the in-line with the dec. axis position for their guidescope, which will avoid this problem, and tube rings are usually designed for this. The "inboard" position would also not be so convenient with a Newtonian main telescope, as it would place the guidescope eyepiece in inconvenient positions, and, in fact, I adopted the standard arrangement for the guidescope on my run-off shed Newtonian.

If a rather substantial subsidiary telescope is required on the same mounting as the main telescope, or, indeed, if two roughly equal-sized telescopes are to be used, then a good solution is the side-by-side dovetail plate or tandem bar, manufactured by Robin Casady and others (Fig. 10.15). This allows the pair of telescopes to be placed an equal distance from the RA axis. They, or their

Figure 10.15. Side-by-side balanced arrangement of two telescopes mounted on a tandem bar (Astro Accessories from RobinCasady.com).

individual dovetail bars, are mounted on a crosswise dovetail bar so that the pair can be adjusted for their centre of gravity to lie in line with the dec. axis. They counterweight each other, and the counterbalancing required on the dec. axis is minimised. This can be extended even to using three telescopes on the same mount, with the use of a "triad bar" (or to more than three if some of the scopes are piggybacked upon one another), or, more reasonably, two telescopes can be combined with a camera, all on the same side-by-side plate.

Cameras are commonly piggybacked on the main telescope, either using the flat top surfaces of the tube rings, or a rail fitted to the telescope or to the rings, and, not being so heavy as telescopes, create less worries as to balance. Long, good-quality telephoto lenses (above 100 mm focal length), combined with DSLRs, can also cover the size range of the large nebulae and clusters effectively, and the lenses can often be obtained for reasonable prices second-hand, as most photographers these days want zoom lenses rather than fixed telephotos. The latter, however, will generally have better optical quality for the money, being simpler. Adaptors can be obtained to fit most lenses made in the past for SLR cameras to modern DSLRs. The most common of the old fittings was the M42 thread, which is the same diameter as the T-thread, but not to be confused with it, as it is coarser. Hence an M42 to DSLR (of whichever make) adaptor is one of the most useful of the bewildering array of adaptors one generally acquires after a short period of dealing with telescopes, cameras and their fittings. When using a camera in this way, the telescope merely acts as a stable, driven platform.

Finder telescopes should have a reasonable aperture, at least 50 mm (2 in.) for an observatory telescope. There is a choice of right-angle and straight-through models. The best ones take standard 31.5 mm (1.25 in.) diagonals and eyepieces, so are interconvertible. The right-angle variant is convenient for Newtonian telescopes, but many Newtonian users like to have a smaller, straight-through finder, or a zero-power pointing device, such as a laser pointer or Telrad finder, mounted on the telescope as well, as it is easiest to get the initial, rough pointing when looking along the telescope tube. The more comfortable right-angle, higher-power finder is then used for precise pointing.

Finder brackets are frequently very poor. Some that are supplied with SCTs do not raise the finder substantially above the main telescope tube, so are obstructed by tube rings holding the main telescope. Newtonians, particularly, require quite high finder brackets, raising the finder at least 6 cm (2.5 in.) above the tube, otherwise it is difficult to get one's eye in the right place. Another piece of unsatisfactory equipment, often supplied, is a bracket that holds the finder with one ring fitted with three setscrews, or with two setscrews and one sprung stud, as one suspension point, the other suspension point being a close-fitting ring fitted with a rubber O-ring. This does not work as well as having two proper rings fitted with setscrews. The rubber ring decays in time. The sprung system is good, as with guidescope mountings. It has become common for finder brackets and laser pointing devices to be fitted to telescopes via a small intermediary dovetail shoe with setscrews, which allows for interchangeability of finders. Unfortunately, the different manufacturers all use different size shoes. Also, the shoe does not provide a very secure fixing base, as it is too small. A wider fixing base, or separate bases for the front and rear rings, is better.

Software

The selection of helpful software available to the amateur astronomer is large and rapidly-advancing. It seems appropriate to mention a few well-regarded titles, though different observers will have their own favourites, and the listing will inevitably go out of date quickly.

While it is possible still to rely on damp, small-scale paper star charts balanced on the knee and illuminated with a faint red torch (probably held in the mouth) while steering the telescope by hand, and while some will argue that this is "the only way to get to know the sky", there is now the alternative of inexpensive and free planetarium software packages, which can be used on their own, or integrated with computer-controlled GOTO mountings. Planetarium software will show you exactly what should be visible from your location at any time and date, and will include much other data on celestial objects. Most astronomical titles are only available for PC, not Mac, unfortunately. This is just a result of the minority niche nature of amateur astronomy, and, possibly, partially a result of the fact that professional astronomers have not tended to employ the Mac platform much. One freeware planetarium package that is available for both platforms is *Stellarium*. This is graphically pretty, but does not do much more than show the basics. For the Mac, *Equinox* is more advanced, and features control capabilities for Meade, other LX200-compatible, and Takahashi mounts.

TheSky from Bisque Software is a well-known commercial application for PC, available in different "levels" depending on the features required, from fairly basic to highly advanced.[2] *TheSky* offers integration with the pointing systems of pretty much all recently-manufactured mountings. This means, in essence, that once the system is fully set-up and the required location data has been entered, and once the telescope has been pointed at a star and that star has been identified to the software (the system has been synchronised), that the software will know exactly where the telescope is pointed in the sky at any time thereafter, provided it is moved using the motors (not by hand). This position will be visible as a marker on the star map on screen. The star map can be viewed at any desired scale, and the number of stars included in the databases of commercial packages is huge, going down, typically, to magnitude 16 (compare with paper star charts of the past, which rarely went below magnitude 8). The software can be used to park the telescope, and a "hibernate" mode on most GOTO mountings ensures that even when all parts of the system are switched off and disconnected, provided the telescope is not moved by hand, at the next startup, the software will still know where in the sky the telescope is pointing. This is invaluable for the observatory owner, who does not want to re-enter data or realign the telescope on the stars at each startup.

It is possible to set up similar mount and planetarium software integration with hand-slewed, non-GOTO mounts, and even with undriven Dobsonians, if they are fitted with axis movement encoders, and an interface that connects to the PC serial port, but the accuracy of such encoders is much lower than that of the higher-quality stepper and servo motor-driven GOTO mounts. The two systems can also be combined to allow software to keep track of the telescope whether it is slewed or hand-pushed; this requires encoders plus servos or steppers. Other well-known software packages integratable with telescope pointing for most mounts are *StarryNight Pro*, *Earth Centred Universe*, *Sky Map Pro*, *SkyTools*, and the free *Cartes du Ciel*.

Software, generally with fewer features, is also available from the major telescope manufacturers, to control only their mounts. Examples are Celestron's *NexRemote*, which only simulates the NexStar hand-controller on the computer screen, and Meade's *Autostar* suite, which displays a comprehensive (though rather unclear) sky map. Astro-Physics mounts work with their *DigitalSky Voice* software, which allows you to command your mount by voice as well as by keyboard, and also with the *PulseGuide* tracking and control software. The system that Meade developed to allow their telescopes to be controlled by computers involves a particular communication language understood by the mount, known as the LX200 protocol. This has been taken up by some other hardware and

---

[2] A very basic version of *TheSky* is currently distributed free with Celestron mounts and telescopes, but without mount-integration features.

software developers as a standard, so a wide range of free and commercial software supports the LX200 protocol, and many non-Meade telescope systems use it, including some home-built ones. The LX200 protocol includes commands for focus control and dome control as well as telescope pointing.

The accuracy of automatic telescope pointing in a perfectly polar-aligned system is normally limited by factors such as lack of mount orthogonality (the axes not being exactly at 90° to one another), instrument flexure, mirror flop (change in collimation) and lack of coincidence of optical and mechanical axes. However, some of these elements have a regular, repeatable quality on a particular permanent set-up, and software exists that can measure these errors for different parts of the sky and compensate for them, giving pointing accuracies of well under one minute of arc, which is nearly always good enough to put an object on even a small (640 × 480 pixel) CCD. *TPoint* from Bisque and *MaxPoint* from Diffraction Limited are packages written to perform this function.

Astro-imagers using CCDs ands DSLRs will need software to capture their images and process them. Other books cover this subject fully,[3] but here are a few brief mentions. In the freeware realm, excellent products are *Registax* (mainly used for processing videos of solar system objects), *DeepSkyStacker*, which does what it says – stacks separate exposures of the same deep-sky object, *IRIS*, which has many image-capture and processing functions hidden within a somewhat non-intuitive design, and *The GIMP*, not an astronomical title *per se*, but a powerful free image-processing package for Linux which works on Mac OS X. In the low-price commercial range is *K3CCDTools*, which is mainly used for capturing and processing deep-sky images taken with long-exposure modified webcams. Advanced astro-imagers working on the deep sky mostly use *Astroart* and *MaxIm DL*, the last being an expensive and comprehensive package of capture and processing tools compatible with a large range of cameras and autoguiders. The best-known general image-processing software on both Mac and PC platforms is Adobe *Photoshop*, which exists in full editions and cut-price versions known as *Photoshop Elements* and *CS2*, which are still very useful to astro-imagers. Some prefer the competing packages *Paintshop Pro* and Corel *PHOTO-PAINT*.

Long-exposure deep sky imaging requires autoguiding for best results, and, in addition to the commercial packages mentioned above that support this, there are two excellent free products that do so, when used with guide cameras and computer interfaces (such as the Shoestring and Astronomiser parallel and USB guide interfaces), or without interfaces, in the case of GOTO mounts. These are *GuideDog*, and the even simpler *PHD Guiding*. Use of these involves downloading the *ASCOM Platform*. ASCOM (which stands for AStronomy Common Object Model) is an initiative to create a comprehensive, mutually-compatible software interface between all sorts of astronomical hardware and software. It includes drivers for mounts, focusers and domes, and much of the software mentioned above relies on it. ASCOM allows user-scripting, that is, the writing of simple programs to co-ordinate the operation of different parts of the observatory, such as telescope-pointing and dome control.

---

[3]See, for example, *The New Amateur Astronomer* by Martin Mobberley, in this series.

The most generally-used features of ASCOM are the mount drivers, available for all the major makes, which automatically allow ASCOM-compliant software to work with most mounts. One of the most useful features of all within ASCOM is the POTH (Plain Old Telescope Handset) driver. The function of this, not terribly obvious from its name, is to create a software "hub", through which multiple programs can control one mount simultaneously. Once the POTH driver is configured to control your telescope, all other ASCOM-compliant software can be configured to use it, rather than the specific driver for your mount, and all should run together harmoniously. The typical use for this is to run autoguiding software at the same time as planetarium software, both controlling one mount. ASCOM is free. It might be commented that, happily, the astronomical software scene is probably more collaborative, and less profit-driven, than almost any other, and this is reflected in the fact that the cost of software is unlikely to be a large part of even the most technically advanced observatory project.

Final Thoughts

There will be as many ways of setting-up a small observatory as there are astronomers, and a book like this can do no more than give a few suggested guidelines that have arisen from the experience of the author and the astronomers he knows, try to give the most practical advice and note some solutions that have worked for some people, and try to stimulate ideas and initiative. Every small observatory is very personal, and different to all others, and that is part of the delight of them. The most important ideas I have tried to stress have been to allow enough space, and to make them uncluttered, safe, and as easy as possible to use. An observatory and all its equipment should be a joy to use, never a discouragement to observing, whether serious or casual. To achieve this usually takes a long time, persistence, patience, and, if we are realistic, not a little money. There are, however, many more expensive and less constructive pursuits.

A practical observatory will repay many times over the time and effort that is put into creating it. The point, ultimately, is the enjoyment and wonder of the heavens, that is the same for an observer in a deckchair on the lawn, counting meteors with the naked eye, as it is for the astronomer in a dome, with a large telescope and the latest electronics, imaging and measuring the stars. Amateur astronomers make a tremendous, and much-underrated, contribution to the culture of mankind. The observatory, of whatever type, is a symbol of the timeless human quest for the appreciation of nature's greatest designs.

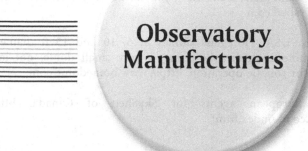

Observatory Manufacturers

Here are details of some observatory manufacturers in Europe, North America and Australia, and a summary of the products they offer, at the time of writing.

Ash-Dome, USA: Metal domes from 2.4 to 9.2 m (8–31 ft) diameter with "up and over" shutters. http://www.ashdome.com

Astro Domes, Australia: Domes from 2.9 to 6.5 m diameter. Electric shutter and automatic dome tracking options. http://www.astrodomes.com

Astro Haven, USA: Manufacture fibreglass "clamshell" domes in sizes 7, 12, 16 and 20 ft diameter, and non-circular ones as well. http://www.astrohaven.com

Alexanders Observatories, England: Build and install run-off roof observatories in timber from 6 ft square to 16 × 10 ft. Also observatory piers. http://www.alexandersobservatories.com

Gambato, Italy: Domes made to measure in steel, aluminium, copper, also sliding roof observatories, and planetarium domes. http://www.gambato.com

Pulsar Optical, England: Fibreglass domed observatories at 7 ft and 9 ft diameter. Will sell domes and tracks separately. http://www.pulsar-optical.co.uk

Sirius Observatories, Australia: Fibreglass observatories, diameters 2.3 m (7.5 ft), 3.5 m (11.5 ft), 6.7 m (22 ft). Motorising and automation options available. http://www.siriusobservatories.com

Sirius Observatories, UK: Agents for the Australian company. http://www.siriusobservatoriesuk.com/

SkyShed, Canada: Build and install run-off roof observatories from 6 to 16 ft^2. Also provide plans, kits and piers; make the SkyShed POD (Personal Observatory Dome). http://www.skyshed.com

Technical Innovations, USA: Fibreglass domes at 6, 10 and 15 ft diameter, motorisation and remote control; robo-dome, a very small dome, not large enough for a person, for robotic operation. http://www.homedome.com

UK Observatories, European agents for SkyShed of Canada. http://www.ukobservatories.co.uk/index.html

Useful Web Links

This lists both commercial and non-commercial sites in order of their subjects' occurrence in this book.

Chapter 1

The ATM site: http://www.atmsite.org/
ATM webring: http://l.webring.com/hub?ring=telescopemaking
ATM group: http://tech.groups.yahoo.com/group/atm_free/
Amateur Telescope Makers' Resource List: http://www.tfn.net/~blombard/
Stellafane ATM pages: http:// www.stellafane.com/atm
Pop Outside Observatory: http://www.popoutside.com
Kendrick Observing Tents: http://www.kendrickastro.com
AstroGazer portable observatory http://www.astrogizmos.com
Sky Window mirrored binsocular mount:
http://www.tricomachine.com/ skywindow/sandt.html
Strathspey binoculars http://www.strathspey.co.uk/
Orion telescopes and binoculars http://www.telescope.com/
QCUIAG: http://tech.groups.yahoo.com/group/QCUIAG/
UK Astro Imaging Forum: http://ukastroimaging.co.uk
Registax software: http://www.astronomie.be/registax/
K3CCDTools software: http://www.pk3.org/Astro/index.htm?k3ccdtools.htm

Chapter 2

BAA Campaign for Dark Skies: http://www.britastro.org/dark-skies/index.html
International Dark-Sky Association: http://www.darksky.org
Information on planning permission in the UK:
http:// www.planningportal.gov.uk

Chapter 3

Scope Armor, US makers of telescope covers: http://scopearmor.com
Group for information on observatory roof and dome automation:
http://tech.groups.yahoo.com/group/Observatory_RoofandDome_Automation/
Group for advice on observatory design and construction:
http://tech.groups.yahoo.com/group/Astronomical_Observatories/
Group dedicated to run-off (roll-off) roof observatories:
http://tech.groups.yahoo.com/group/roll_off_roof/
Amateur Astronomical Observatories: list of links to web articles on a large
number of observatories: http://obs.nineplanets.org/obs/obslist.html

Chapter 7

The Webb Society - sell very nice laminated star-charts:
http:// www.webbsociety.org

Chapter 8

Mel Bartels: products and designs for customised telescope control (US):
http://www.bbastrodesigns.com/
AWR Technology: drive and computer control systems, including retrofitting of
old mounts (UK):
http://www.awrtech.co.uk
JMI Telescopes (Jim's Mobile Inc.): focus control systems:
http:// www.jimsmobile.com
Technical Innovations: Robo-Focus: http://www.homedome.com
Switched Systems (UK): focus motors: http://www.switchedsystems.co.uk

Chapter 9

Les Granges: http://www.lesgrangesastro.talktalk.net/
Dave Tyler: http://www.david-tyler.com
Damian Peach: http://www.damianpeach.com

Chapter 10

Howie Glatter's laser collimation products, including Barlowed collimators and collimation plugs: http://www.collimator.com

Aberrator software: http://aberrator.astronomy.net/

Astronomy Boy has instructions for servicing some common mountings: http://www.astronomyboy.com

Astrozap dew shields: http://www.astrozap.com

Parallax Instruments: Tube rings for all telescopes: http:// www.parallaxinstruments.com

Andy Homeyer: Heavy-duty tube cradles and dovetail systems: http://www.intint.com/andy/index.html

Robin Casady: Telescope dovetail systems, multiple telescope mounting systems, counterweights: http://www.robincasady.com

Comprehensive listing of planetarium software available for different platforms: http://astro.nineplanets.org/astrosoftware.html

Stellarium Open-source planetarium software: http://www.stellarium.org

Equinox planetarium and telescope-control software for Mac: http://www.microprojects.ca/

Software Bisque: *TheSky* and associated software, also manufacture the Paramount ME mounting: http://www.bisque.com

Starry Night Pro planetarium and telescope-control software: http://shop.avanquest.com/

Earth *Centered Universe* planetarium and telescope-control software: http:// www.nova-astro.com

SkyMap Pro planetarium and telescope-control software: http://www.skymap.com

SkyTools observing software: http://www.skyhound.com

Cartes du Ciel free planetarium and telescope-control software: http://www.ap-i.net/skychart/index.php

Diffraction Limited: *MaxPoint* software for telescope mount modelling and *MaxIm DL* software for image acquisition and processing: http://www.cyanogen.com

DeepSkyStacker image processing software http://deepskystacker.free.fr/english/index.html

IRIS image acquisition and processing software: http://www.astrosurf.com/buil/ us/iris/iris.htm

Astroart image acquisition and processing software: http://www.msb-astroart.com/

Astronomiser: Astro-imaging equipment including guide interfaces: http://www.astronomiser.co.uk/

Shoestring Astronomy: astro-imaging accessories including guide interfaces: http://www.store.shoestringastronomy.com/

GuideDog autoguiding software: http:www.barkosoftware.com

PHD Guiding software: http://www.stark-labs.com/

ASCOM: Standards initiative for astronomical software/hardware communication: http://ascom-standards.org/index.html

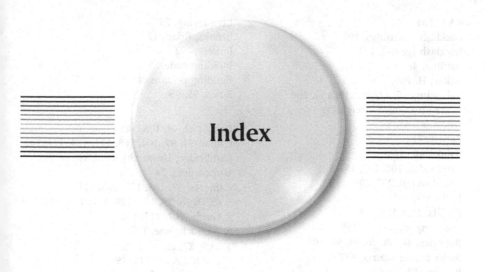

Index

Other Titles in this Series

(Continued from page ii)